George K. Stegman
Jerry E. Jenkins

Engineering Technology Department
Western Michigan University
Kalamazoo, Michigan

Series 3
First Edition

engineering graphics problems

Macmillan Publishing Company
New York

Collier Macmillan Publishers
London

Printed in the United States of America

MACMILLAN PUBLISHING COMPANY
866 Third Avenue, New York, New York 10022

COLLIER MACMILLAN CANADA, LTD.

ISBN 0-02-416310-4

Printing: 3 4 5 6 7 8 Year: 8 9 0 1 2 3

ISBN 0-02-416310-4

preface

Engineering Graphics Problems—Series 3 First Edition, is designed to be used with *Engineering Graphics, Third Edition* by Giesecke, Mitchell, Spencer, Hill, Loving and Dygdon. All reference material is directed to that textbook. Consequently, the topic organization is divided into three areas: Part One "Technical Drawing"; Part Two, "Descriptive Geometry"; and Part Three, "Graphs and Computer Graphics." This workbook, may however, be used in conjunction with other suitable reference textbooks at the instructor's discretion.

Engineering Graphics Problems—Series 3, has been brought together from the author's industrial and educational experiences, and has been evaluated and articulated to meet the current needs of the engineering graphics student. Because we believe that a thorough understanding of graphical fundamentals is essential, added emphasis has been placed on the areas of multiview projection, sectional views, auxiliary views and dimensioning. Additionally, the area of graphical representation has been reduced to facilitate a more streamlined approach to this area.

Due to the fast-growing and widespread use of the computer in graphical disciplines, the authors have taken a fundamental, yet functional, approach to the section on computer graphics. We believe that if students are to become literate in the area of computer graphics, they must initially be able to understand the graphical language and acquire the ability to think spatially. Without these capabilities, efforts in the area of computer graphics can be frustrating. Therefore, a supplementary approach has been taken to Chapter 30, introducing the student to line and point orientation in coordinate systems. The student is given an additional opportunity, using generic command language, to create short subroutines to plot two- and three-dimensional objects without the actual use of a CAD system.

A feature, worthy of note, is the inclusion of three-dimensional instructional aids designed to be used with the topics of multiview projection, sectional views, and the auxiliary view of a line. These sheets can be removed from the workbook and folded to produce a three-dimensional aid to the visualizing process.

All problems are printed on 8.5″ x 11.0″ sheets in conformity with American National Standards for engineering drawing practices. This size facilitates handling and filing by the instructor and student. An increased number of problem sheets have been printed on vellum, an aspect many instructors have been looking forward to for some time.

Our special thanks to James Fox, who supplied a number of problems on geometric constructions. Additional thanks are due, as well, to the many industries that have contributed ideas and problem material. For the most part, these materials have been revised to meet the needs of the engineering graphics classroom. Comments, ideas, and suggestions from users of this workbook are welcomed and appreciated by the authors.

Our deepest appreciation to our wives, Lois and Dianne, for their many hours of patience and understanding.

George K. Stegman
Jerry E. Jenkins
Engineering Technology Department
Western Michigan University
Kalamazoo, Michigan

contents: topics and drawings

instructions

References are to *Engineering Graphics,* Third Edition, by Giesecke, Mitchell, Spencer, Hill, Loving, and Dygdon.

It should be noted this workbook frequently uses metric and inch dimensions that are not exact conversional equivalents. The equivalent of 25.4mm = 1.00″ has often been rounded to 25mm = 1.00″. To simplify the conversion process, the following conversions may also be used: 12 or 12.5mm = .50″, 6mm = .25″, 3mm = .12″, etc. Where critical calculations are required, the exact equivalent should be used.

Part One. Technical Drawing

In general, the authors recommend the following leads for use in technical drawing: for guide lines and construction lines, a sharp 4H lead. For center lines, section lines, extension lines, and dimension lines, a sharp 2H lead. For object lines and lettering, a sharp H or F lead.

Drawing 1. Vertical Capitals and Numerals. References: §§3.1 to 3.18. Letter the indicated characters in the spaces provided using an HB or F lead. All lettering should be **black** with sharp edges. Draw light horizontal guide lines from the starting marks under COURSE, SECT., DRAWN BY, and DATE. Use random vertical quide lines as necessary and letter these sections with the appropriate information as assigned by your instructor.

Drawing 2. Vertical Lowercase Lettering. References: §§3.21, 3.22, 3.24. In the upper area of the sheet, letter the indicated characters in the spaces provided using a sharp HB or F lead. At the bottom of the sheet, construct three sets of horizontal guide lines to duplicate the three lines of lettering shown. All lettering should be **black** with clean edges.

Drawing 3. Vertical Capitals and Numerals. References: §§3.1 to 3.18, 3.24. In area l of the sheet, draw horizontal guides as assigned by your instructor and letter the text assigned to area 1 below. Random vertical guide lines may be drawn to assist in keeping the lettering vertical. Repeat this procedure for areas 2-6 using the assigned text for each area below. Area 6 may be used as a test if the instructor so desires.

Area 1

Properly proportioned letters are only part of quality lettering. The spacing between letters is of equal significance. If the space between each letter appears equal, the word will be easy to read. This is also true for numerals. 1 2 13 24 35 67 89 10.

Area 2

To prevent hand fatigue and cramping, move the entire hand while forming each stroke of the letter or numeral. Entire hand movement relieves the muscle tension between the thumb, index, and middle finger.

Area 3

Softer grade leads are better for lettering such as the HB, F and H The pencil should have a rounded conical point. The uniformity of the point can be maintained by slightly rotating the pencil after each stroke.

Area 4

The engineer, designer, and drafter employ two guide lines equal to the height of the capital letter. The ANSI standard height of letters for notes and dimensions is 0.12″. Horizontal guide lines are always used.

Area 5

Frequently, letters must be slightly expanded or condensed in a word or group of words to fit into a given space or area. When this technique is used, legibility must not be sacrificed.

Area 6

Today many industries prefer vertical single-stroke gothic style lettering. Other industries allow either vertical or inclined lettering, the choice being made by the individual.

Drawing 4. Vertical Capitals and Numerals. References: §§3.10, 3.14 to 3.18, 3.20, 3.24, 11.5, 11.7 to 11.11, 11.14 to 11.16. On the upper area of the sheet are shown a number of lettering applications. Below each, reproduce the lettering and arrowheads using a sharp HB or F lead. In the lower area of the sheet, letter the months on the graph as indicated in the completed illustration. Center the drill note about the given center line. Also letter the given shop notes below the example. The height of the drill and shop notes is given in the parentheses. Use guide lines and make all lettering **black** with sharp edges.

Drawing 5. Inclined Capitals and Numerals. References: §§3.1 to 3.17, 3.19. Letter the indicated characters in the spaces provided using an HB or F lead. All lettering should be **black** with sharp edges.

Drawing 6. Inclined Lowercase Lettering. References: §§3.21, 3.23, 3.24. In the upper area of the sheet, letter the indicated characters in the spaces provided using a sharp HB or F lead. At the bottom of the sheet, construct three sets of horizontal guide lines to duplicate the three lines of lettering shown. All lettering should be **black** with sharp edges.

Drawing 7. Inclined Capitals and Numerals. References: §§3.1 to 3.17, 3.19, 3.24. In area 1 of the sheet, draw horizontal guide lines as assigned by your instructor and letter the text assigned in area 1 below. Random guide lines may be drawn to assist in keeping the lettering inclined. Repeat this procedure for areas 2-6 using the assigned text for each area below. Area 6 may be used as a test if the instructor so desires.

Area 1

Properly proportioned letters are only part of quality lettering. The spacing between letters is of equal significance. If the space between each letter appears equal, the word will be easy to read. This is also true for 1 2 13 24 35 67 89 10.

Area 2

To prevent hand fatigue and cramping, move the entire hand while forming each stroke of theletter or numeral. Entire hand movement relieves the muscle tension between the thumb, index, and middle finger.

Area 3

Softer grade leads are better for lettering such as the HB, F and H. The pencil should have a rounded conical point. The uniformity of the point can be maintained by slightly rotating the pencil after each stroke.

Area 4

The engineer, designer, and drafter employ two guide lines equal to the height of the capital letter. The ANSI standard height of letters for notes and dimensions is 0.12″. Horizontal guide lines are always used.

Area 5

Frequently, letters must be slightly expanded or condensed in a word or group of words to fit into a given space or area. When this technique is used, legibility must not be sacrificed.

Area 6

Today many industries prefer single-stroke gothic style lettering. Other industries allow either vertical or inclined lettering, the choice being made by the individual.

Drawing 8. Inclined Capitals and Numerals. References: §§3.10, 3.14 to 3.18, 3.19, 3.20, 3.24, 11.5, 11.7 to 11.11, 11.14 to 11.16. On the upper area of the sheet are shown a number of lettering applications. Below each, reproduce the lettering and arrowheads using a sharp HB or F lead. In the lower area of the sheet, letter the months on the graph as indicated in the completed illustration. Center the drill note about the given center line. Also letter the given shop notes below the example. The height of the drill and shop notes is given in the parentheses. Use guide lines and make all lettering **black** with sharp edges.

Drawing 9. Layout of Lines. References: §§2.1 to 2.25, 2.28, 2.29, 2.33, 2.47. Draw lines to fill the required areas as follows:

Area 1. Set of visible lines, 45° with horizontal, 3/16″ apart from the given line.

Area 2. Set of hidden lines, 30° with the horizontal, 10 mm apart from the given line.

Area 3. Set of section lines, 60° with the horizontal, 15 mm apart from the given line.

Area 4. Set of vertical center lines spaced ¼″ apart. Locate the center of the area by drawing light diagonals from each corner. From the centerpoint, draw the first vertical center line. Step off on each side of this line ¼″ divisions and draw the required lines.

Area 5. Set of dimension lines, 15 mm apart and perpendicular to given line. See §2.33.

Area 6. Set of horizontal visible lines spaced ⅜″ apart. Locate the center of area as in area 4. Through the centerpoint draw the first horizontal line. Step off on each side of this line ⅜″ divisions and draw the required lines.

Drawing 10. Scale Lengths (Inch Units). References: §§2.25 to 2.33.

Area 1. Use the architect's or engineer's scale as necessary. Draw lines A-H to the specified length using the indicated scales.

Area 2. Using an architect's or engineer's scale as necessary, measure lines J-P according to the indicated scale. Draw guide lines as needed and letter neatly the measured length in the space provided along each line.

Drawing 11. Scale Lengths (Metric Units.) References: §§2.26, 2.32, 2.33.

Area 1. Using a metric scale, draw lines A-H to the specified length using the indicated scales.

Area 2. Using a metric scale, measure lines J-P according to the indicated scale. Draw guide lines as needed and letter neatly the measured length in the space provided along each line.

Drawing 12. Geometric Constructions. References: §§4.1 to 4.7, 4.12, 4.14, 4.17, 4.18, 4.32, 4.52. Show light construction on lines on all problems.

Area 1. Reference: §4.18. Draw a perpendicular line as indicated.

Area 2. References: §§4.11, 4.12. Construct the indicated angle.

Area 3. Reference: §4.17. Divide the line into the indicated proportions.

Area 4. Reference: § 4.32. Draw an arc (partial circle) through the indicated points.

Area 5. Reference: §4.52. Complete the ellipse as described using the concentric circle method.

Area 6. Reference: §4.14. Draw a line parallel to the curved line at the required distance.

Drawing 13. Geometric Constructions. References: §§4.1 to 4.7, 4.21, 4.36, 4.39, 4.40, 4.41, 4.42.

Area 1. Complete the full-scale layout of the BELLCRANK. (Note starting point A.)

Area 2. Complete the full-scale layout of the TRIPARM. (Note starting point A.)

Drawing 14. Geometric Constructions. References: §§4.1 to 4.7, 4.21, 4.40, 4.41, 4.42.

Area 1. Complete the full-scale layout of the LENS TEMPLATE.

Area 2. Complete the full-scale layout of the INVOLUTE EXHAUST. This is based on constructing the involute of a line. See Reference: §4.65(a). (Note starting point A.) Line AB = .50 inches.

Drawing 15. Geometric Constructions. References: §§4.1 to 4.7, 4.21, 4.37, 4.39, 4.41, 4.42. Complete the full-scale layout of the CHAINHOOK. (Note starting point A.)

Drawing 16. Geometric Constructions. References: §§4.12. Make a full-scale, single-view drawing of the AIRSHUTER, which is designed to control the volume of air intake on an air duct. The screw slot is to be designed to allow the same amount of travel as required by the air intake slot. Allow 0.7 mm clearance between the metric machine screw and the slot itself. Use guide lines and letter the title below the drawing as well as the material callout. (See Appendix 15 for the maximum body diameter of the metric machine screw.)

Drawing 17. Technical Sketching. References: §§5.1 to 5.10.

Area 1. Sketch the object on the enlarged grid as indicated. Make all final lines **black** so they will stand out from the grid lines.

Area 2. Sketch the object on the enlarged grid system as indicated. Make all final lines **black** so they will stand out from the grid lines.

Multiview Teaching Aid. References: §§6.1 to 6.4. Remove the sheet from the workbook. Fold according to the directions as an aid in visualizing multiview projection and the principal planes of projection.

Drawing 18. Multiview Technical Sketching. References: §§5.1 to 5.10, 5.18 to 5.31. Sketch the six views of the BASE as indicated. Space the six views two squares apart. Make all lines **black** so they will stand out from the grid lines.

Drawing 19. Multiview Technical Sketching. References: §§5.1 to 5.10, 5.18 to 5.31. Sketch the missing view of each object in the indicated spaces. Make all final lines **black** so they will stand out from the grid lines.

Drawing 20. Multiview Projection. References: §§5.25, 5.26, 6.1 to 6.20. In each problem, two complete views are given. Complete the missing third view as indicated using instruments.

Drawing 21. Multiview Projection. References: §§6.1 to 6.20, 6.27 to 6.30, 6.33. In each problem two complete views have been given. Complete the missing third view as indicated using instruments.

Drawing 22. Multiview Projection. References: §§6.1 to 6.20, 6.27 to 6.30, 6.33. Complete the missing third view in each problem as indicated. Two completed views are given. Use instruments.

Drawing 23. Multiview Projection. References: §§6.1 to 6.20, 6.27 to 6.30, 6.33, 6.34, 6.35. In each problem two completed views are given. Add the third view as indicated using instruments.

Drawing 24. Multiview Projection. References: §§6.1 to 6.30, 6.33, 6.34, 6.35. In each problem two completed views are given. Complete the missing view as indicated using instruments.

Drawing 25. Multiview Projection. References: §§6.1 to 6.26. In each problem two completed views are drawn. Complete the missing third view as indicated. Each problem deals primarily with oblique surfaces in multiview projection. Use instruments.

Drawing 26. Multiview Projection. References: §§6.1 to 6.30, 6.33. Complete the missing third view in each problem. Two completed views are drawn.

Drawing 27. Multiview Projection. References: §§6.1 to 6.30, 6.33. Add the missing third view in each problem. Two completed views are given.

Sectional View Teaching Aid. References: §§7.1 to 7.8. Remove the sheet from the workbook. Fold according to the instructions as an aid in visualizing full and half sections.

Drawing 28. Full and Half Sections. References: §§6.34, 7.1 to 7.7, 7.15. Sketch the full and half sections as indicated. Sketch the section lines thin to contrast with the visible lines. All final lines should be clean-cut and **black** to stand out from the grid lines.

Drawing 29. Sectional Views. References: §§2.11, 6.34, 6.35, 7.1 to 7.8, 7.11, 7.15. Draw the indicated sections using instruments. In all four problems draw the cutting plane in its proper position. Use the general-use symbol for all sectioned areas.

Drawing 30. Full Sectional Views. References: §§7.1 to 7.6, 7.12, 7.15. Draw the indicated sections in each of the six problems. Use the general-use symbol for all sectioned areas. In area 3, the web is 0.625″ deep and is to be centered in the sectional view.

Drawing 31. Full Section. References: §§7.1 to 7.6, 7.12. Make a full sectional view of the ADJUSTING BASE using instruments. The cutting plane is to pass horizontally through the center of the 1.03″ reamed hole in the front view.

Drawing 32. Aligned Sections. References: §§7.1 to 7.8, 7.12, 7.13.

Area 1. Complete the aligned section as a half section as indicated by the cutting plane line. Draw a revolved section of the rib on the unsectioned portion of the side view. Add missing centerline in front view.

Area 2. Draw the aligned section as indicated. If assigned, add finish marks to the given and sectioned views.

Drawing 33. Removed Sections. References: §§7.1 to 7.7, 7.10, 7.15. Draw removed sections A-A, B-B, C-C, and view D-D of the VALVE SHAFT in the indicated positions. Properly identify each removed sectional view. Use general-use section lines.

Drawing 34. Primary Auxiliary Views. References: §§8.1 to 8.10. Sketch the required auxiliary views indicated in each problem. Make all final lines clean-cut and **black** to stand out from the grid lines.

Drawing 35. Primary Auxiliary Views. References: §§8.1 to 8.10, 8.16. Complete the indicated auxiliary view in each problem. Problem 1 requires a complete auxiliary as does Problem 3. Problems 2 and 4 require only partial auxiliary views to show the indicated surfaces.

Drawing 36. Primary Auxiliary Views. References: §§8.1 to 8.10, 8.12. Complete the indicated auxiliary view of each problem. Complete the plotted curve with the help of an irregular curve.

Drawing 37. Primary Auxiliary Views. References: §§8.1 to 8.10, 8.12, 8.24. Draw the indicated auxiliary view for each problem.

Drawing 38. Primary Auxiliary Views. References: §§8.1 to 8.10, 8.16. Complete the entire auxiliary view of each problem. Identify the surface(s) that are projected in their true size and shape with a TSS.

Drawing 39. Secondary Auxiliary Views. References: §§8.11, 8.19 to 8.22, 11.14. Draw the necessary partial auxiliary views to show the true size and shape of oblique surface A. Dimension the dihedral angle that is formed by the two plates in the appropriate view.

Drawing 40. Secondary Auxiliary Views. References: §§8.13, 8.19 to 8.24. Draw the necessary auxiliary views of the DOVETAIL SUPPORT to show the true size and shape of oblique surface A. Locate the true position of the four center lines (for drilled holes) in the true size and shape view. Draw an additional auxiliary view to show the true size and shape of the dovetail slide. (Dimensions are given in the detail of the dovetail slide.) Then, using reverse construction, draw the dovetail slide to complete the top and front views.

Drawing 41. Revolutions. References: §§8.13, 9.1 to 9.11.

Area 1. Revolve line 1-2 about axis 3-4. Using reverse construction, locate the new position of line 1-2 in the front and profile views. Dimension the angle of rotation in the auxiliary view.

Area 2. Revolve the V-BLOCK about point A until the bottom of the milled V-slot is in a vertical position. Draw the top, front, and right-side views in the revolved position. Dimension the true angle of the V-slot in the top view.

Drawing 42. Isometric Sketching. References: §§5.1 to 5.8, 5.12 to 5.18. Using an F lead, complete the isometric sketches as indicated. Note the given starting point A for each sketch. Omit hidden lines, Make all final lines clean-cut and **black** to stand out from the grid lines. (Note inclined surface on problem 2).

Drawing 43. Isometric Drawing. References: §§16.4 to 16.19. Complete the isometric drawings as indicated for each problem. Note the starting point A for each drawing. Do not transfer distances directly with dividers, since the drawings are not full size. Use the given dimensions. Unless otherwise instructed, use the approximate four center ellipse method for the hole in Problem 1. All construction lines should be drawn lightly with a 4H lead and should not be erased unless otherwise instructed. Omit all hidden lines.

Drawing 44. Isometric Drawing. References: §§16.4 to 16.19. Complete the isometric drawings for each problem. Note the starting point A for each drawing. Use the approximate four center ellipse method on both problems unless otherwise instructed. Make all construction lines with a 4H lead and do not erase unless instructed. Omit all hidden lines.

Drawing 45. Isometric Drawing. References: §§16.4 to 16.9. Complete the isometric drawing of the OFFSET GUIDE. Note the starting corner indicated A. Use the approximate four center ellipse method unless otherwise instructed. Use a sharp 4H lead in laying out all construction lines and do not erase unless instructed. Omit hidden lines.

Drawing 46. Oblique Sketching. References: §§5.4, 5.5, 5.15, 5.16, 17.13. Make freehand sketches of the parts shown using the starting corners indicated. Make all final lines clean-cut and **black** to stand out from the grid lines.

Drawing 47. Oblique Drawing. References: §§17.1 to 17.6, 17.10. Complete the oblique pictorials using instruments. Both problems are to be cabinet pictorials. Note the starting points for each problem. Study Fig. 17.13 carefully before starting the FAUCET MOCK-UP. The SUPPORT PAD is to be drawn as an oblique section. The cutting plane is to pass vertically through the center of the two in-line holes. Use the general-use section symbol. Omit hidden lines.

Drawing 48. Dimensioning. References: §§11.1 to 11.26, 11.29, 11.30, 11.31. Use the complete decimal dimensioning system with decimal inch values. Add dimensions mechanically spacing the dimension lines as assigned by your instructor.

Space. 1 The small holes are drilled. The two holes on the left are spaced 45° from the vertical center line. The large hole is counterbored. See Fig. 11.42.

Space 2. The hole is bored. Use the note .625—.615 BORE. All fillets and rounds are .06 R.

Drawing 49. Dimensioning. References: §§11.1 to 11.26, 11.32 to 11.35. Add dimensions mechanically using the decimal dimensioning system with metric values to describe the NEEDLE CORE. Use the note M16 x 1.5-4g6g for the external thread. The scale is double size and Detail A is drawn to a scale of 4X. Space dimension lines as noted by your instructor. Dimension the rounded end of the part on the enlarged detail.

Drawing 50. Dimensioning. References: §§11.1 to 11.26, 11.28 11.31, 11.32, 12.1. Using the decimal dimensioning system with metric values, add dimensions mechanically to describe the VISE JAW. Use the thread note M12 x

1.75-6H for the internal thread. Since the part is cast iron, add finish marks to the milled rectangular slot on the end and to the inside of the larger rectangular opening on the top. Radii are taken from the noted centerpoints. The 10 mm reamed hole has a tolerance of + 0.05 mm, -0.00 mm. The overall, general tolerance is ± .025 mm. Space dimension lines as noted by your instructor.

Drawing 51. Dimensioning. References: §§11.1 to 11.26, 11.31. Add dimensions to the SADDLE BRACKET using the decimal dimensioning system with decimal-inch equivalents. Use only two-place decimals and locate all holes from center lines and finished surfaces. The large hole is drill and spotfaced. The three smaller holes are drilled. The overall general tolerance is +.010, -.010. Note that the material is cast iron and the quantity is 31. Space dimension lines as noted by your instructor. Complete finish marks.

Drawing 52. Dimensioning. References: §§11.1 to 11.26, 11.31, 12.2. Draw the necessary views of the WASHER SEAL as indicated. Add dimensions noting the limits of size. Express the limit dimensions in fractional form (one over the other). Add the required notes and finish marks and identify any needed sections.

Drawing 53. Dimensioning. References: §§12.1 to 12.11. Using one of the columns below designated by your instructor, dimension each problem according to the specification. All of the dimensions given in the table are nominal or basic sizes for each problem. Convert the basic sizes to the full decimal equivalent. Apply all tolerances to the full decimal value and then round off to four decimal places. Express all limits in fractional form with one limit above the other. Use Appendixes 5-9 to determine the limits of fits for Problems 1-6. Use Fig. 12.11 in applying the proper tolerance to each sizing operation in Problems 7-12. (The broached hole is cylindrical.)

	A	B	C	D
1.	3/8	13/16	1 27/32	2 3/16
2.	15/16	2 1/8	3 7/8	5 35/64
3.	17/32	4 19/32	2 13/16	3 7/8
4.	6 3/16	3 15/16	4 17/32	5 3/8
5.	2 15/32	5 5/32	3 37/64	4 9/16
6.	2 31/32	2 39/64	5 5/16	1 11/32
7.	1 1/4	3/8	1 1/4	2 23/32
8.	3/4	1 1/4	2 3/8	6 45/64
9.	2 1/4	15/16	1 1/2	2 5/8
10.	1 15/16	7/8	2 3/4	2 9/16
11.	6 1/8	4 1/2	5 11/16	4 7/16
12.	10 17/64	6 5/8	7 19/32	5 15/64

Drawing 54. Dimensioning. References: §§11.1 to 11.26, 11.31, 12.1 to 12.7. Dimension the SHAFT EXTENSION COLLAR using the decimal dimensioning system with decimal-inch equivalents. The 1.00″ hole is to be toleranced for an RC4 fit. All dimensions are to be two-place decimals with a general tolerance of +0.010, -0.010. The four holes are reamed with a tolerance of +0.001, -0.001 and are equally spaced. All rounds and fillets are 0.09R. The webs are also equally spaced. The material is type 333 aluminum. The quantity is 131.

Drawing 55. Dimensioning. References: §§11.1 to 11.26, 11.31, 11.34, 11.42, 12.1 to 12.10. Add dimensions to the ALIGNMENT PLATE using the decimal dimensioning system with decimal-inch equivalents. The part is cast from grade 30 cast iron. The tolerance on all "as cast" (unfinished) surfaces is +0.01, -0.01. The same tolerance applies to machined two-place decimal values. All fillets and rounds are 0.06R. Additionally, all burrs and sharp edges are to be removed.

The largest finished O.D. has a basic size of 3 5/16″ and is turned on the lathe for an RC1 fit. The second largest O.D. is to be ground to a basic size of 1 1/32″ with an RC4 fit. The smallest finished I.D. has a basic size of 3/4″ with a tolerance of +0.004, -0.001. The enlarged end of this hole is turned with an average roughness height that should be selected and dimensioned.

All 12 holes are spaced 30° apart. The 3 holes spaced 120° apart are drilled and reamed on a 1.625″ circle. The tolerance on this diameter is +0.002, -0.002. The remaining three sets of holes are placed on a 2.500″ diameter circle. The tolerance on this circle is +0.004, -0.000. These drilled holes are 0.31″ in diameter.

Letter the necessary notes. The quantity required is 150.

Drawing 56. Dimensioning. References: §§11.1 to 11.26, 11.31, 11.36, 12.1. Make a detail drawing of the CASTER FORK on a separate sheet(s) of vellum. Use the directions on the problem sheet in adding dimensions to the drawing. Letter the notes required.

Drawing 57. Threads and Fasteners. References: §§13.1 to 13.11, 13.26, 13.30, 13.31, 13.34, 13.35. Draw the assigned fasteners as indicated. Refer to Appendixes 13, 15, 18, 21, 22, and 24 for information on hexagon nuts, machine screws, woodruff keys, plain washers, lock washers, and taper pins, respectively. See Fig. 13.34 for information on set screws. The taper pin and set screw have alternate positions on the mandrel. Draw the alternate hole for the taper pin and show the alternate drill spot to seat the set screw. Complete the section lines around all features.

Drawing 58. Threads and Fasteners. References: §§13.1 to 13.11, 13.19, 13.21 to 13.25, 13.26, 13.29 to 13.32. Draw the indicated fasteners and thread representations. Show the tap drill hole if the fastener requires a tap drill. Complete all section lines around the fasteners. (The effective length of the 1/4-NPT is 0.502.)

Part Two. Descriptive Geometry

Accuracy and Layout—accuracy is of prime importance in solving graphical problems. Sharp 2H and F leads are recommended for most line work. A sharp 4H lead is recommended for guide lines and construction work. Measurements should be taken from the center of lines.

The basic scalar unit is the millimeter unless otherwise noted. Therefore, a scale of 1:1000 is noted as being 1 mm = 1000 mm or 1 mm = 1 meter. It is suggested for accuracy that the student identify all points and lines in the constructed view(s) with the correct notation. It is also helpful to label folding lines in such view(s). Dimension angles, distances, bearing, grades, true size, etc., on the view in which they are measured.

Drawing 59. Visibility Of Points and Lines. References: §§19.1, 19.2. Use the standard alphabet of lines (2.11) to complete the solutions. In area 5 the piercing point of the lines with the planes is given. In area 6 the line of intersection is provided between the intersecting planes.

Auxiliary View Teaching Aid. References: §§8.1 to 8.4, 19.4, 19.6. Remove the sheet from the workbook and fold according to the directions at the bottom of the sheet. The horizontal, frontal, and profile views of line ab can be viewed in three-dimensional form. The bearing will be shown in the horizontal plane. Also shown, will be the true length and slope of line ab on the auxiliary plane.

Drawing 60. True Length by Auxiliary View. References: §§19.4 to 19.7. Use auxiliary views to solve all problems. In area 2 dimension the angle in the proper view.

Drawing 61. True Length by Revolution. References: §§9.10, 19.5. Use American National Standard phantom lines for all lines in the revolved position. In area 1 dimension the true length, bearing, and slope on the appropriate view(s). In area 2 dimension the angles in the proper view(s).

Drawing 62. Cones of Revolution. References: §§9.10, 19.5. Use revolution to provide solutions to all problems.

Drawing 63. True Size of Plane. References: §§6.15, 8.1, 8.21, 19.10, 19.11. Use secondary auxiliary views to solve all problems. In areas 1 and 2 show values and calculations for the area. Record the area to two-place decimals on the view. In area 3 dimension the angle to cut the railing on the auxiliary view.

Drawing 64. Piercing Points—Edge-View Method. Reference: §19.12. Use the edge-view method on this sheet and show the proper visibility in all views. Identify the edge view (EV) where appropriate and circle the piercing points in all views.

Drawing 65. Piercing Points-Given-View Method. Reference: §19.12. Use given-view method (cutting plane method) to solve all problems. Label EV (edge view) on the lines used as a cutting plane and circle the piercing points in all views. In area 3 label the line of intersection (LI) where appropriate and shade the visible portion of plane 1-2-3.

Drawing 66. Dihedral Angle. References: §§8.11, 19.10. Dimension the required angles for all problems on the proper view. Show complete visibility where appropriate. In area 3 determine the angle between the planes by extending the edges of the planes until they meet in the auxiliary view.

Drawing 67. Angle Between Line and Plane. Reference: §19.14. Use successive auxiliary views and revolutions where convenient to solve the problem. Dimension the angle on the appropriate view.

Drawing 68. Parallelism. References: §§19.6, 20.1 to 20.5. In area 1 dimension the bearing and true length on the drawing.

Drawing 69. Perpendicularity. References: §§20.5 to 20.7, 20.9.

Areas 1 and 2. These problems are not to be solved with additional views.

Areas 3 and 4. These problems may be solved with an auxiliary view. In area 4, dimension the shortest distance where appropriate on the view.

Drawing 70. Skew Lines. References: §§19.6, 20.9. These problems are designed to be solved by the point view method. Due to space limitations other methods are not recommended.

Area 1. Dimension the shortest distance on the drawing. By reverse construction show this line in the top and front view.

Area 2. Dimension the bearing, true length, and slope of the tunnel on the appropriate view.

Drawing 71. Intersections of Curves and Planes. Reference: § 21.3. Established the line of intersection between the plane and curved surface in area 1. In area 2 establish the line of intersection between the conical structure and the plane.

Drawing 72. Solid-Solid Intersection. References: §§21.4, 21.5, 21.6, 21.8. Establish the figure of intersection on all problems. Show hidden lines and determine the proper visibility of each member.

Drawing 73. Intersections of Prisms. References: §§21.5, 21.8. Establish the figure of intersection on both problems. Show all hidden lines and determine the proper visibility for each prism.

Drawing 74. Intersection of Cones. References: §§21.7, 21.8, 21.10. Establish the figure of intersection between the cones. Show hidden lines and determine the proper visibility for each cone.

Drawing 75. Parallel-Line Development. References: §§22.1 to 22.5. Show full developments inside up. Omit bases.

Space 2. Calculate the length of the development.

Drawing 76. Radial-Line Development. References: §§22.6, 22.7, 22.8. Show developments as indicated. Omit bases.

Drawing 77. Triangulation. References: §§22.10 to 22.14. Develop the transition pieces as indicated. Note starting point for each development. Omit bases.

Drawing 78. Mining and Geology. References: §§24.3 to 24.6. Dimension the strike and dip where appropriate on the proper view in each problem. In area 3 use dot-dash coding for outcrop lines. These should be heavy lines. Section the outcrop lines.

Drawing 79. Civil Engineering-Cut and Fill. Reference: §24.6. Show light construction on all plotted points. Identify the cut and fill areas by cross-hatching in opposite directions, shading with colored pencil or other means.

Drawing 80. Spherical Geometry. Reference: §24.7. Dimension and identify all required data on the drawing as well as in the spaces provided. Plot the arcs of the triangle on both views of the sphere if assigned by your instructor.

Drawing 81. Spherical Naviagtion. Reference: §24.8. Dimension and identify the required data on the appropriate views. Show calculations for the required great circle distance in nautical miles. Plot the course in both views if assigned by your instructor.

Drawing 82. Concurrent Vector Analysis. References: §§25.1, 25.2, 25.3, 25.5, 25.6. Dimension all scalar quantities on each vector diagram.

Drawing 83. Nonconcurrent Coplaner Vectors. Reference: §25.4. Solve for the resultant forces for the loaded beams. Begin the string diargams at A. Use the space directly below each beam to construct the funicular diagram. Indicate the resultant forces for each beam on the appropriate drawing.

Drawing 84. Nonconcurrent Coplaner Vector. Reference: §25.4. Dimension and identify the resultant forces acting on the Fink truss. Start the stress diagram at A. Use the area marked Force analysis/sketches for any needed graphical "thinking."

Part Three. Graphs and Computer Graphics

Drawing 85. Rectangular Coordinate Graph. References: §§26.1, 26.2, 26.5, 26.6, 26.7. Plot the following data on the given rectangular coordinate graph. Use recommended point symbols to establish the data points on the graph. Establish the graphical curves with the aid of an irregular curve. Letter appropriate captions and scales. Letter the title on the graph in the space provided.

Grease Replacement Periods for
Bearings at 150°F (65.5°C)

Grease Renewal Periods (hours)	*Bearing Speed Factors*		
	Taper and Spherical Roller	*Plain Roller*	*Ball*
10			15
7			46
5			74
4		0	93
2.5		31	131
2	0	45	151
1	45	92	210
5	92	140	270
1	200	250	

Drawing 86. Semilogarithmic Graph. Reference: §26.9. Plot the data used in Drawing 85 on the given semilogarithmic coordinate graph. Use recommended point symbols to establish the data points. Letter the appropriate captions and scales. Letter the title on the graph in the space provided.

Drawing 87. Logarithmic Coordinate Graph. Reference: § 26.10. Plot the following data on the given three-cycle logarithmic graph. Use recommended point symbols to establish the data points. Letter appropriate captions, scales, and title on the sheet.

Cost and Quantity Relation Among Die Casting,
Sand Casting, and Mechanical Machining

	Relative Cost		
Quantity	*Die Casting*	*Machining*	*Sand Casting*
10	250	32	25
100	15	16	6.5
1000	3.5	10	5
10,000	1.3	9	5

Introduction to Computer Graphics: References: §§30.1 to 30.14. The computer graphics section has been developed in conjunction with Chapter 30 of *Engineering Graphics*, Third Edition, by Giesecke, Mitchell, Spencer, Hill, Loving, and Dygdon. However, additional information is provided in the "Introduction to Computer Graphics" to help provide solutions to the specific problem material contained in Drawings 88-91.

Drawing 88. Two-Dimensional Coordinate Plots. References: §§30.1 to 30.14.

Area 1. Using the dimensional information given, calculate the X, Y coordinates for each lettered point. (*Note:* Use the tangent of 60° to calculate the X, Y values for point F.)

Area 2. Using the dimensional information given, calculate the X, Y coordinate values for the indicated points. (*Note:* Use the tangent of 30° to calculate the X, Y values for point G.) Then, using the commands for two-dimensional drawings, develop a subroutine that will plot the drawing in a clockwise direction.

Drawing 89. Two-Dimensional Coordinate Plot. References: §§30.1 to 30.14. Write a subroutine, using the commands for two-dimensional drawings, that will plot a full-size drawing of the SHAFT GUIDE. Keep in mind that any Y *values below the origin* and any X *values to the left of the origin* are *negative*. Indicate any negative values in the written subroutine with a negative sign (-). Develop the drawing in a clockwise direction.

Drawing 90. Three-Dimensional Coordinate Plot. References: §§30.1 to 30.14. Determine X, Y, and Z coordinate values for points A-M. Then, develop a subroutine using the commands for three-dimensional drawings to plot the BEVEL BASE. Check your subroutine for accuracy of point values and DRAW commands.

Drawing 91. Three-Dimensional Coordinate Plot. References: §§30.1 to 30.14. Using instruments and a scale, plot the three-dimensional object defined by the listed subroutine. Check your plot for accuracy and label all important points. Indicate, with arrows, the direction each line was plotted. Also number which line was plotted first, second, third, etc.

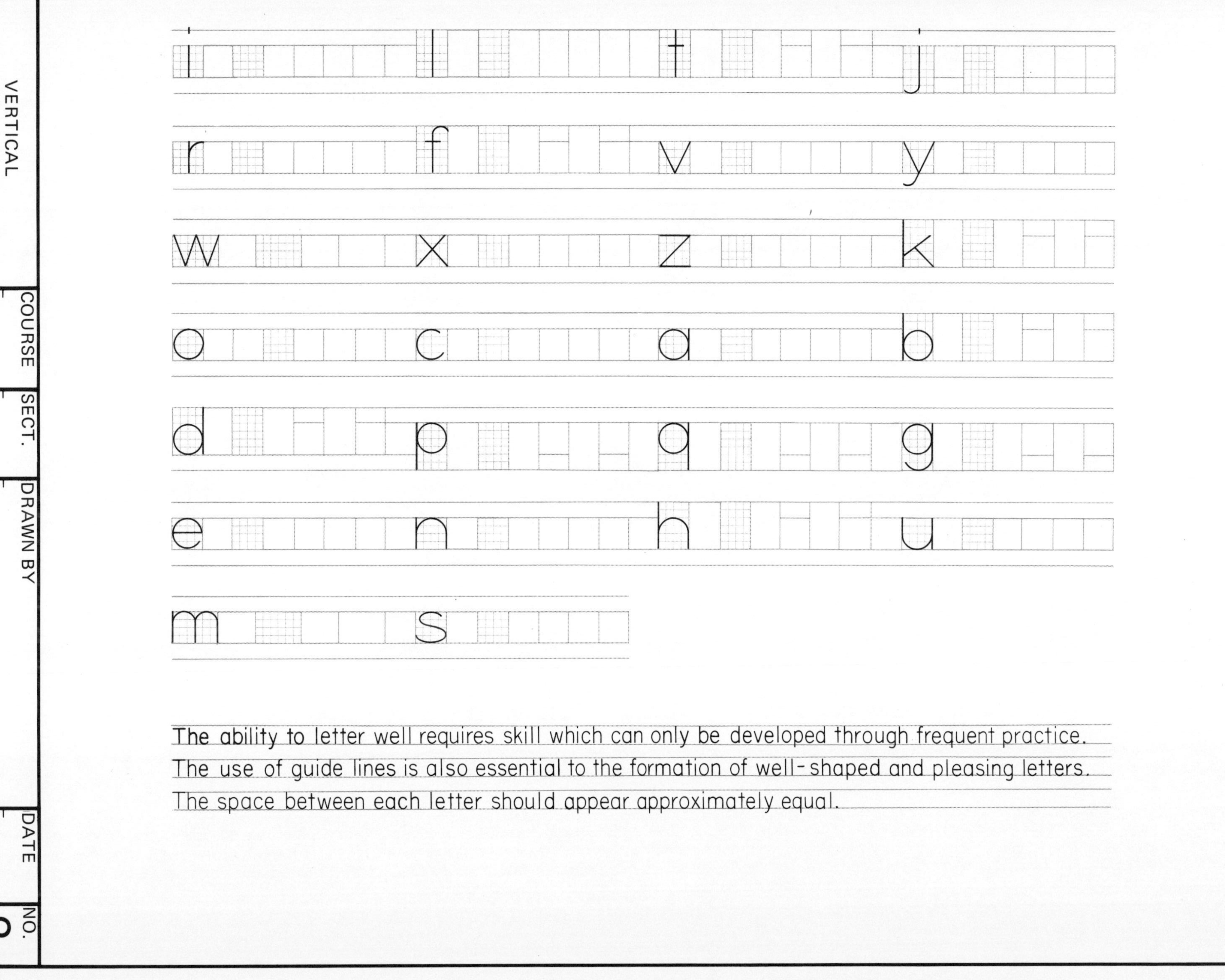
i l t j
r f v y
w x z k
o c a b
d p q g
e n h u
m s
The ability to letter well requires skill which can only be developed through frequent practice.
The use of guide lines is also essential to the formation of well-shaped and pleasing letters.
The space between each letter should appear approximately equal.
VERTICAL LETTERING
COURSE
SECT.
DRAWN BY
DATE
NO.
2

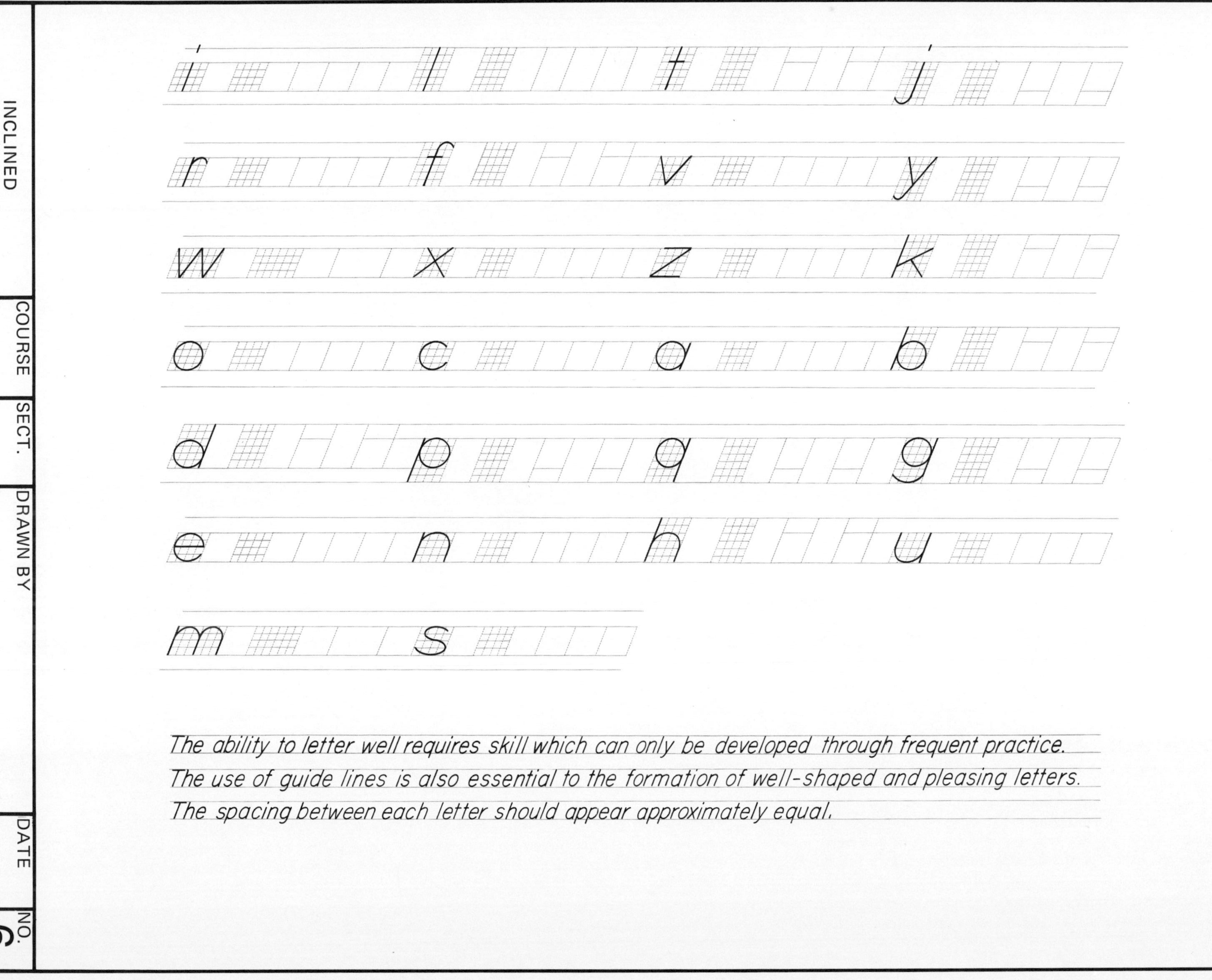

INCLINED LETTERING	COURSE	SECT.	DRAWN BY	DATE	NO. 6

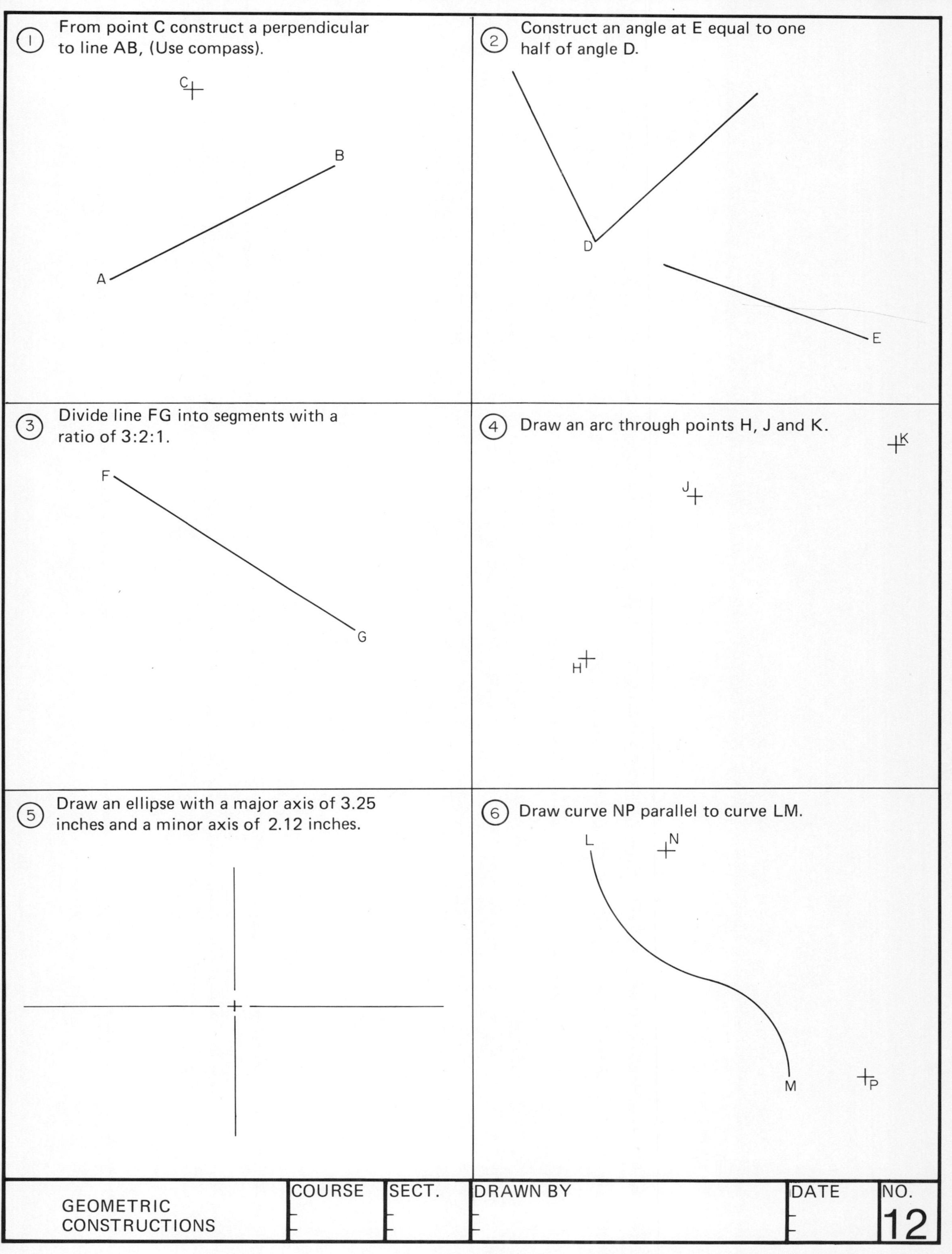

(1) From point C construct a perpendicular to line AB, (Use compass).
C
B
A
(2) Construct an angle at E equal to one half of angle D.
D
E
(3) Divide line FG into segments with a ratio of 3:2:1.
F
G
(4) Draw an arc through points H, J and K.
K
J
H
(5) Draw an ellipse with a major axis of 3.25 inches and a minor axis of 2.12 inches.
(6) Draw curve NP parallel to curve LM.
L
N
M
P
GEOMETRIC CONSTRUCTIONS
COURSE
SECT.
DRAWN BY
DATE
NO.
12

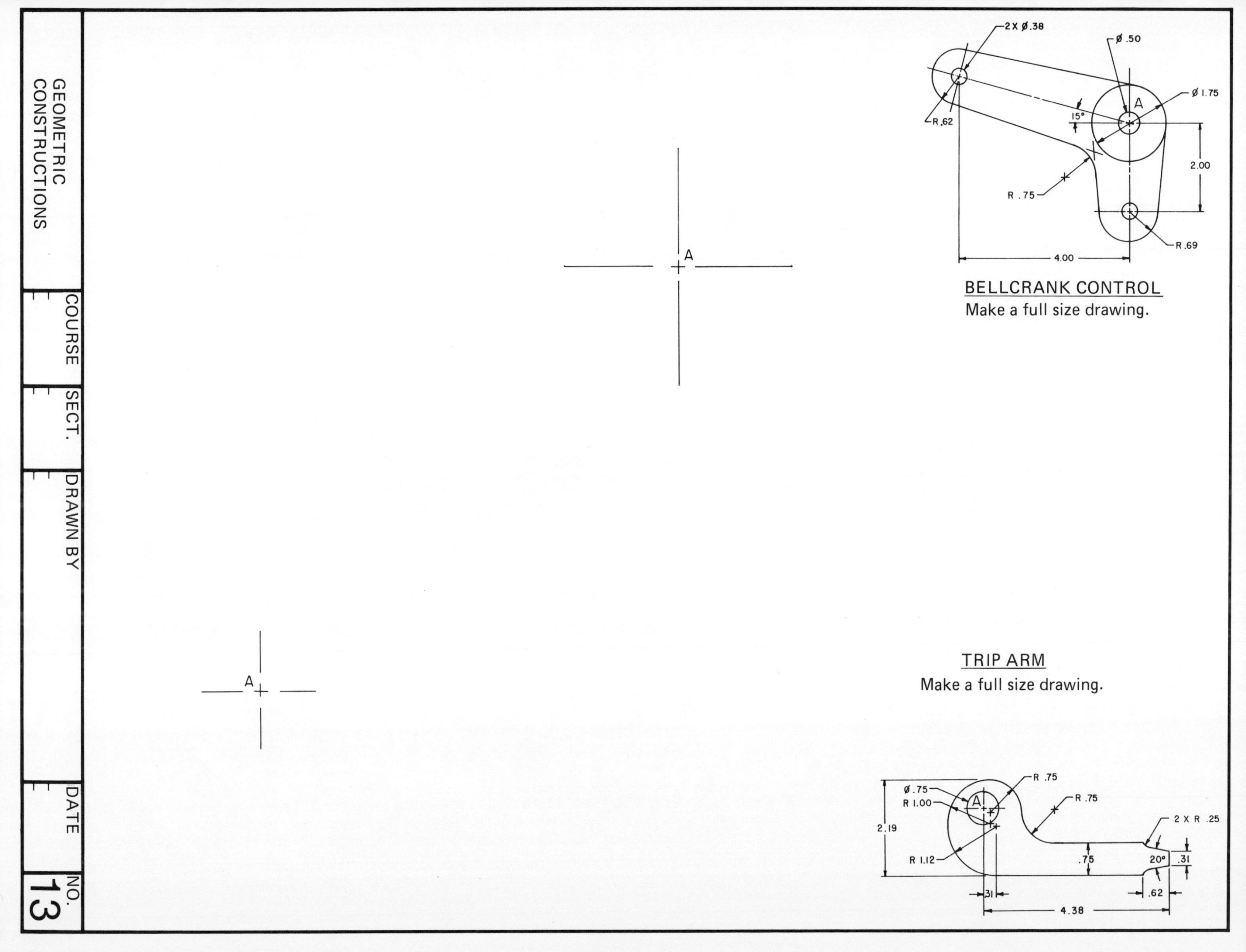

2X Ø .38
Ø .50
Ø 1.75
A
15°
R .62
2.00
R .75
R .69
4.00
BELLCRANK CONTROL
Make a full size drawing.
A
A
TRIP ARM
Make a full size drawing.
R .75
Ø .75
R 1.00
A
R .75
2 X R .25
2.19
R 1.12
.75
20°
.31
.31
.62
4.38
GEOMETRIC
CONSTRUCTIONS
COURSE
SECT.
DRAWN BY
DATE
NO.
13

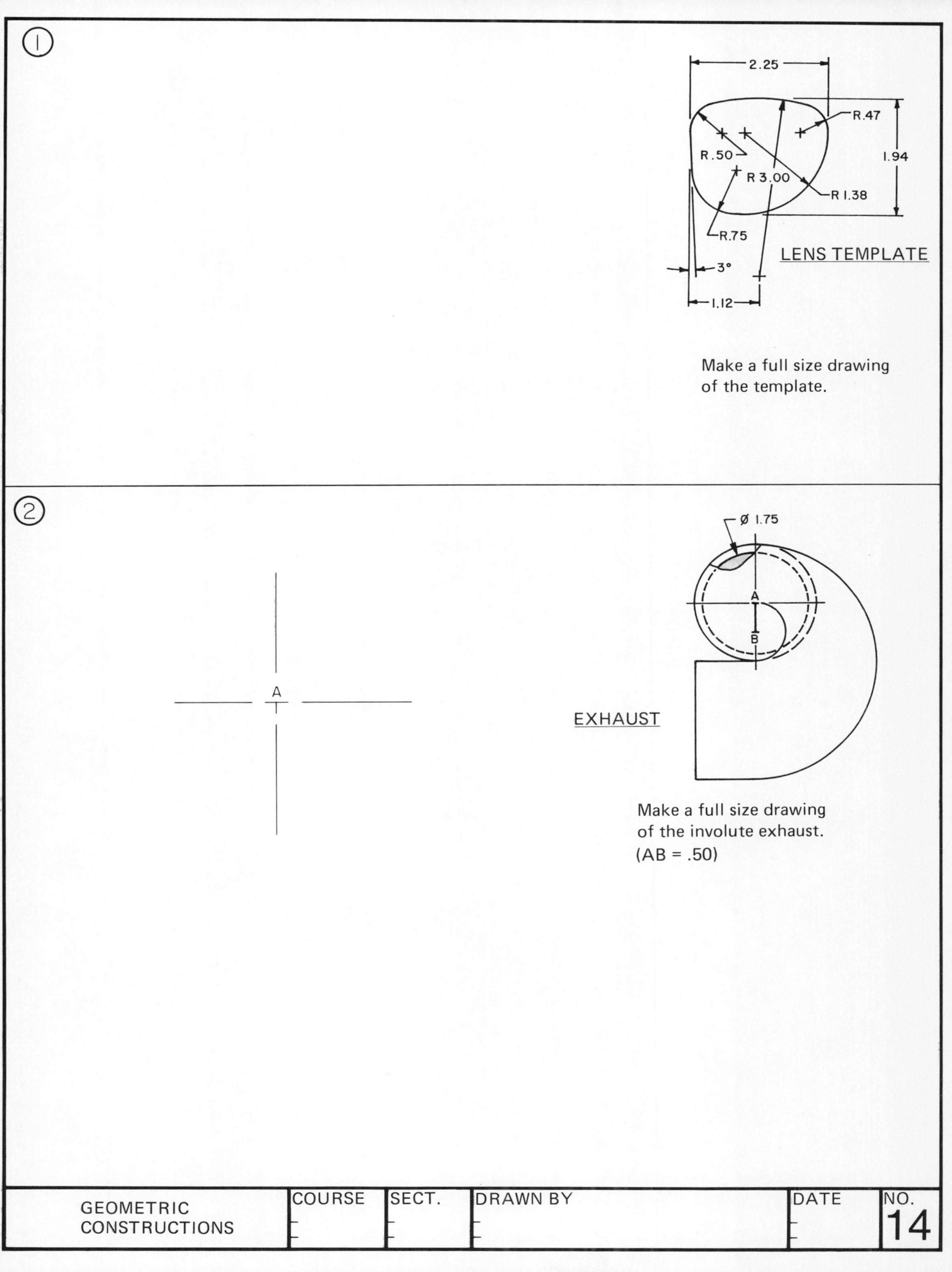
①
2.25
R.47
1.94
R.50
R 3.00
R 1.38
R.75
3°
1.12
LENS TEMPLATE
Make a full size drawing
of the template.
②
Ø 1.75
A
B
A
EXHAUST
Make a full size drawing
of the involute exhaust.
(AB = .50)
GEOMETRIC
CONSTRUCTIONS
COURSE
SECT.
DRAWN BY
DATE
NO.
14

MULTIVIEW PROJECTION TEACHING AID

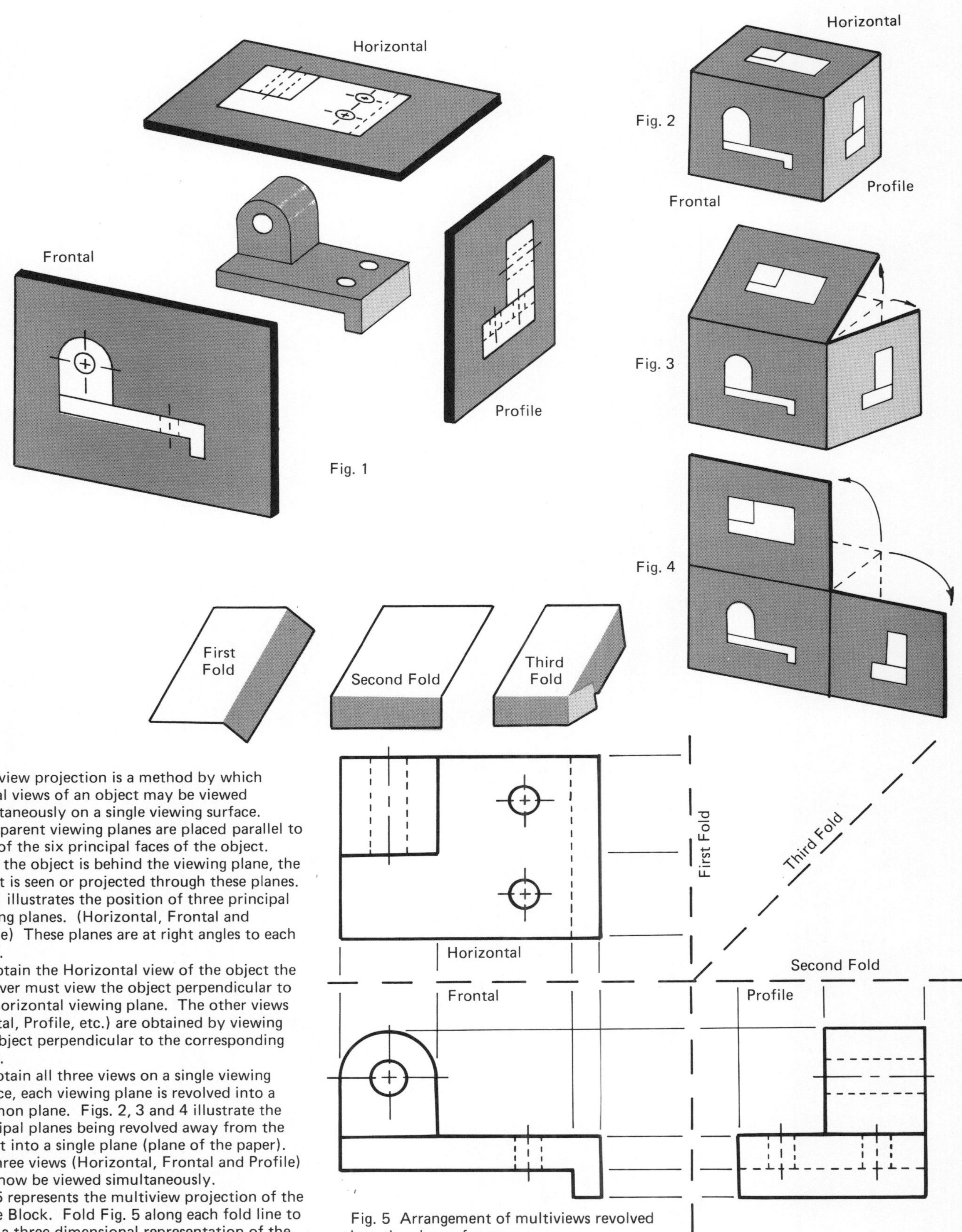

Fig. 5 Arrangement of multiviews revolved into the plane of paper.

Multiview projection is a method by which several views of an object may be viewed simultaneously on a single viewing surface. Transparent viewing planes are placed parallel to each of the six principal faces of the object. Since the object is behind the viewing plane, the object is seen or projected through these planes. Fig. 1 illustrates the position of three principal viewing planes. (Horizontal, Frontal and Profile) These planes are at right angles to each other.

To obtain the Horizontal view of the object the observer must view the object perpendicular to the Horizontal viewing plane. The other views Frontal, Profile, etc.) are obtained by viewing the object perpendicular to the corresponding plane.

To obtain all three views on a single viewing surface, each viewing plane is revolved into a common plane. Figs. 2, 3 and 4 illustrate the principal planes being revolved away from the object into a single plane (plane of the paper). All three views (Horizontal, Frontal and Profile) may now be viewed simultaneously.

Fig. 5 represents the multiview projection of the Hinge Block. Fold Fig. 5 along each fold line to form a three-dimensional representation of the three viewing planes.

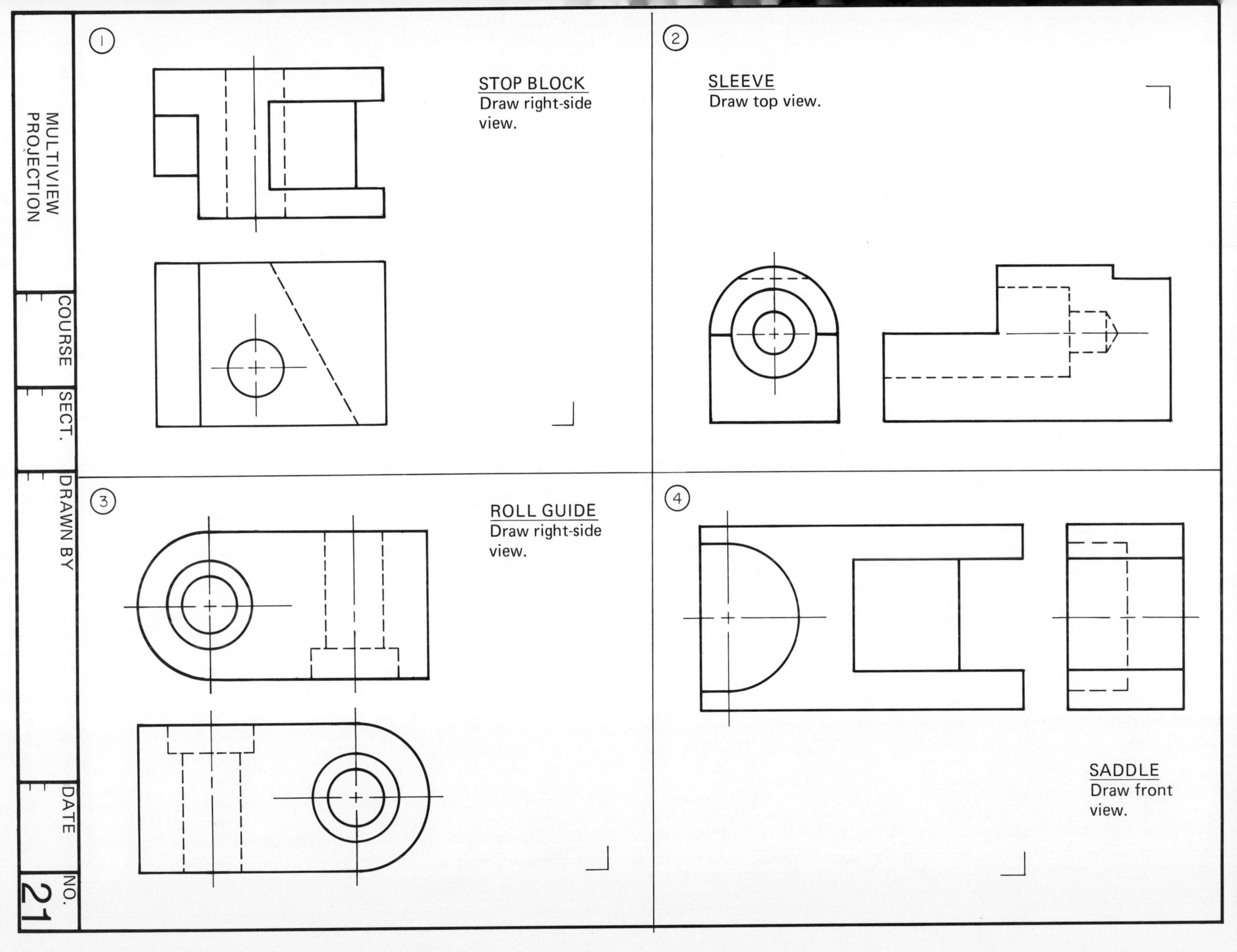

1
STOP BLOCK
Draw right-side view.
2
SLEEVE
Draw top view.
3
ROLL GUIDE
Draw right-side view.
4
SADDLE
Draw front view.
MULTIVIEW PROJECTION
COURSE
SECT.
DRAWN BY
DATE
NO.
21

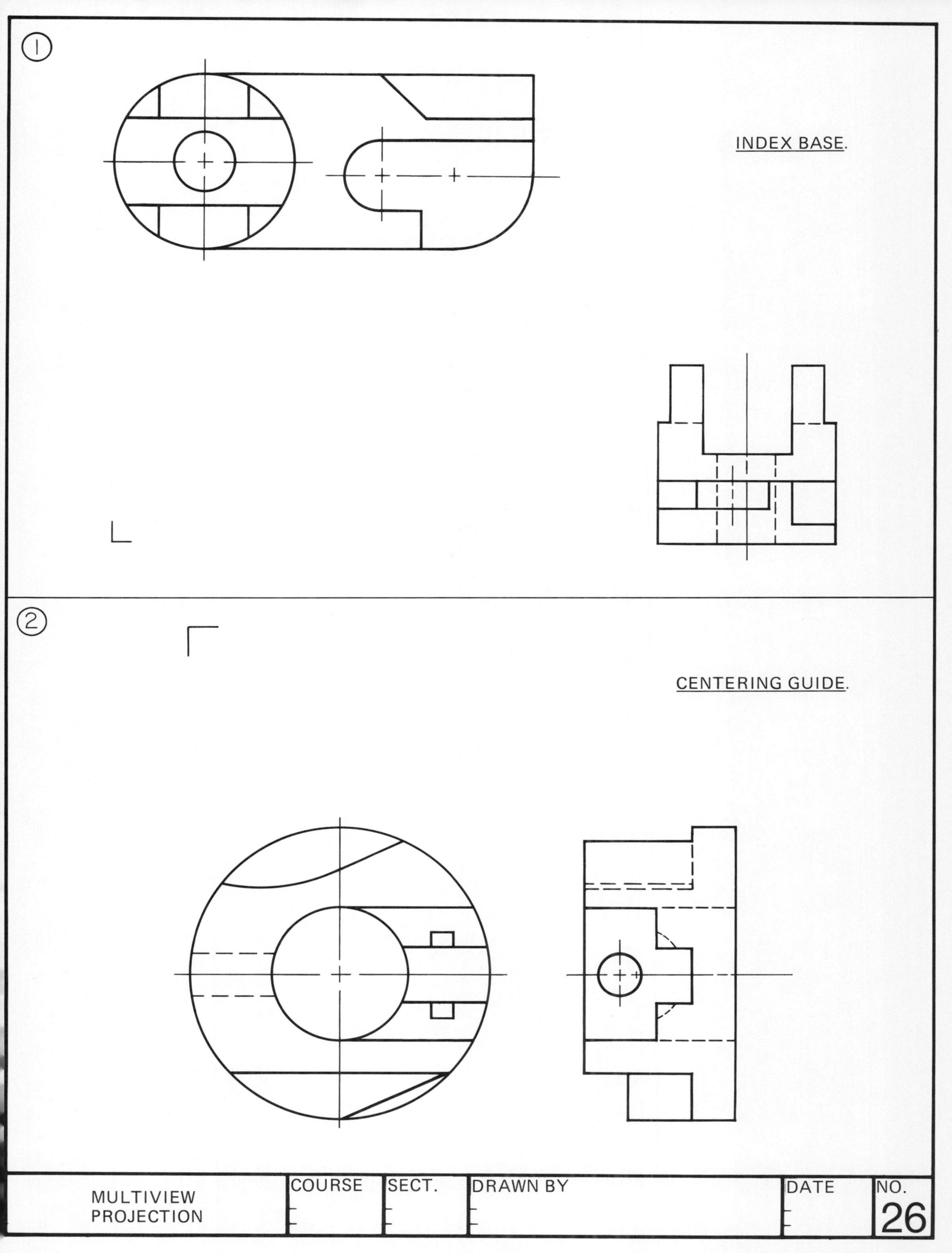
1
INDEX BASE.
2
CENTERING GUIDE.
MULTIVIEW
PROJECTION
COURSE
SECT.
DRAWN BY
DATE
NO.
26

SECTIONAL VIEW TEACHING AID

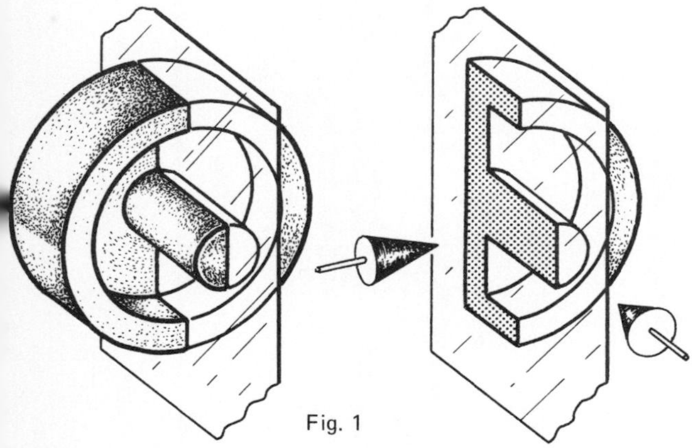

Fig. 1

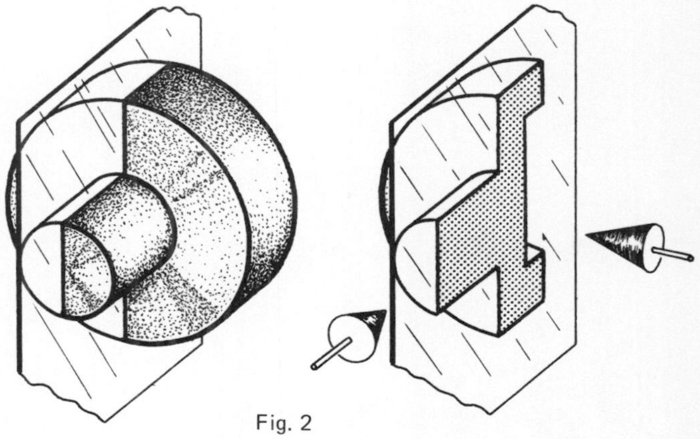

Fig. 2

To understand how sectional views are drawn, place the cutting plane through the object. By removing the portion between your eye and the cutting plane, the interior detail of the object is revealed.

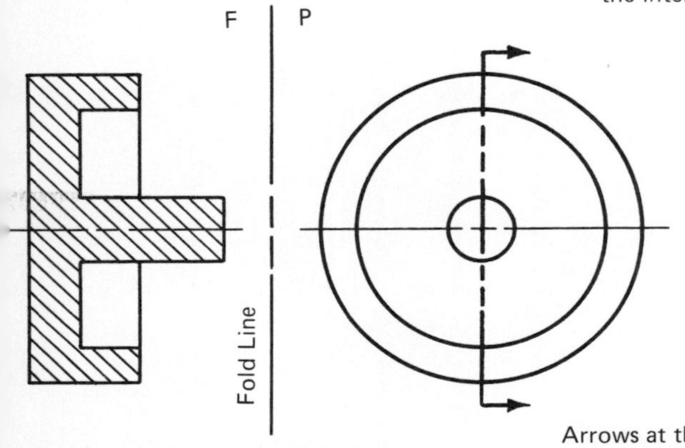

F | P

Fold Line

Fig. 1A Full Section

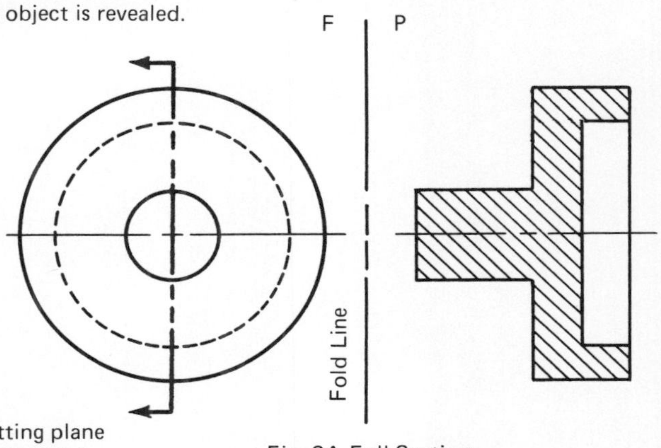

F | P

Fold Line

Fig. 2A Full Section

Arrows at the end of the cutting plane indicate the direction of sight. Fold views along fold lines. Area adjacent to fold line represents plane through which interior of object is viewed.

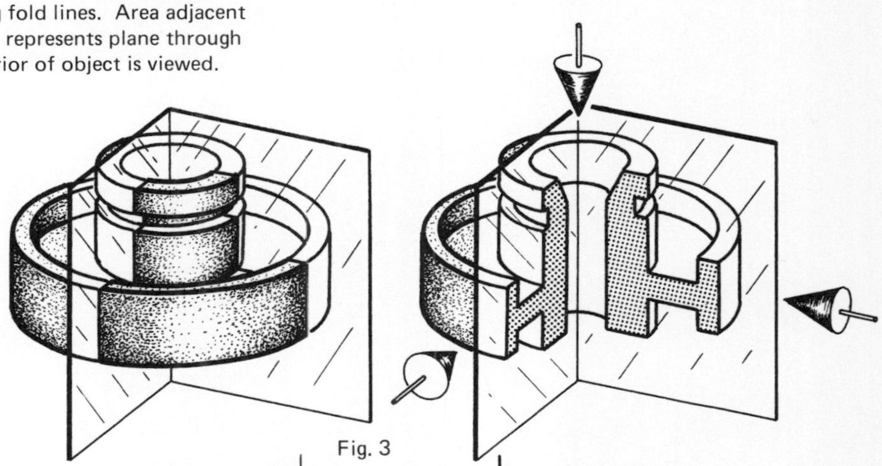

Fig. 3

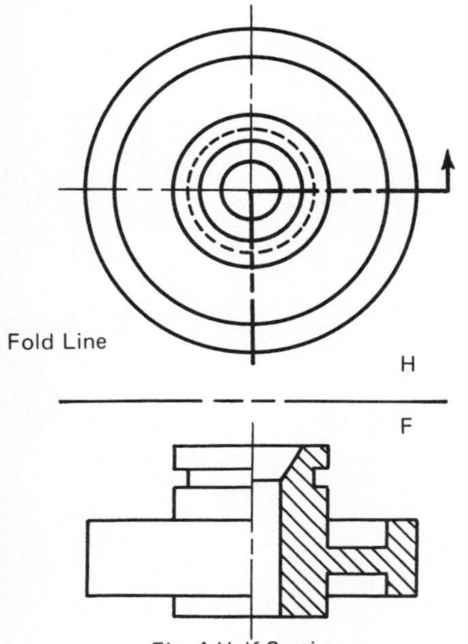

Fold Line

H

F

Fig. 4 Half Section

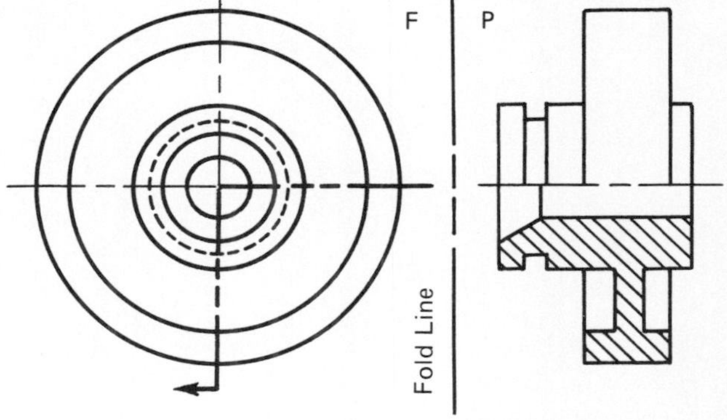

F | P

Fold Line

Fig. 5 Half Section

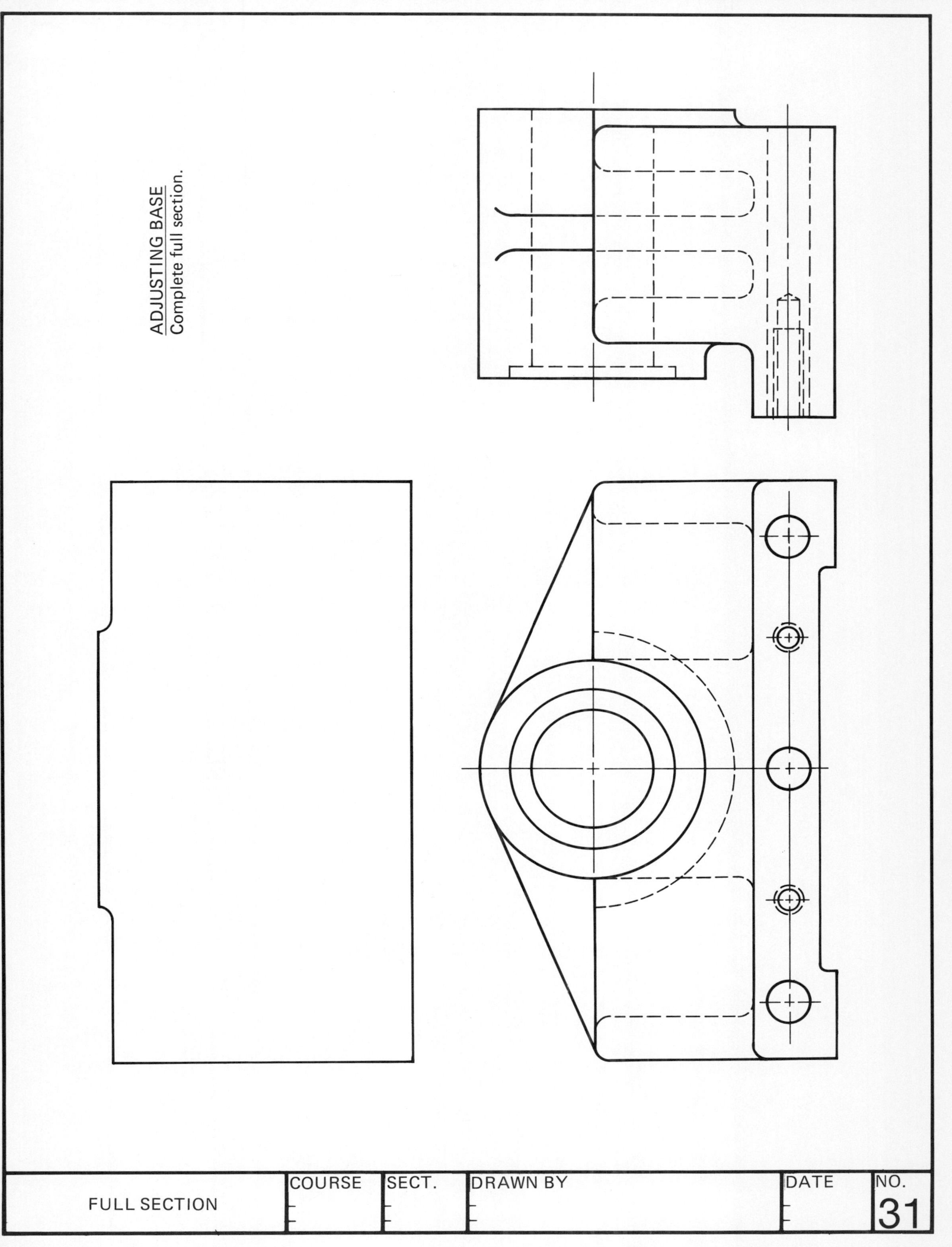
ADJUSTING BASE
Complete full section.
FULL SECTION
COURSE
SECT.
DRAWN BY
DATE
NO.
31

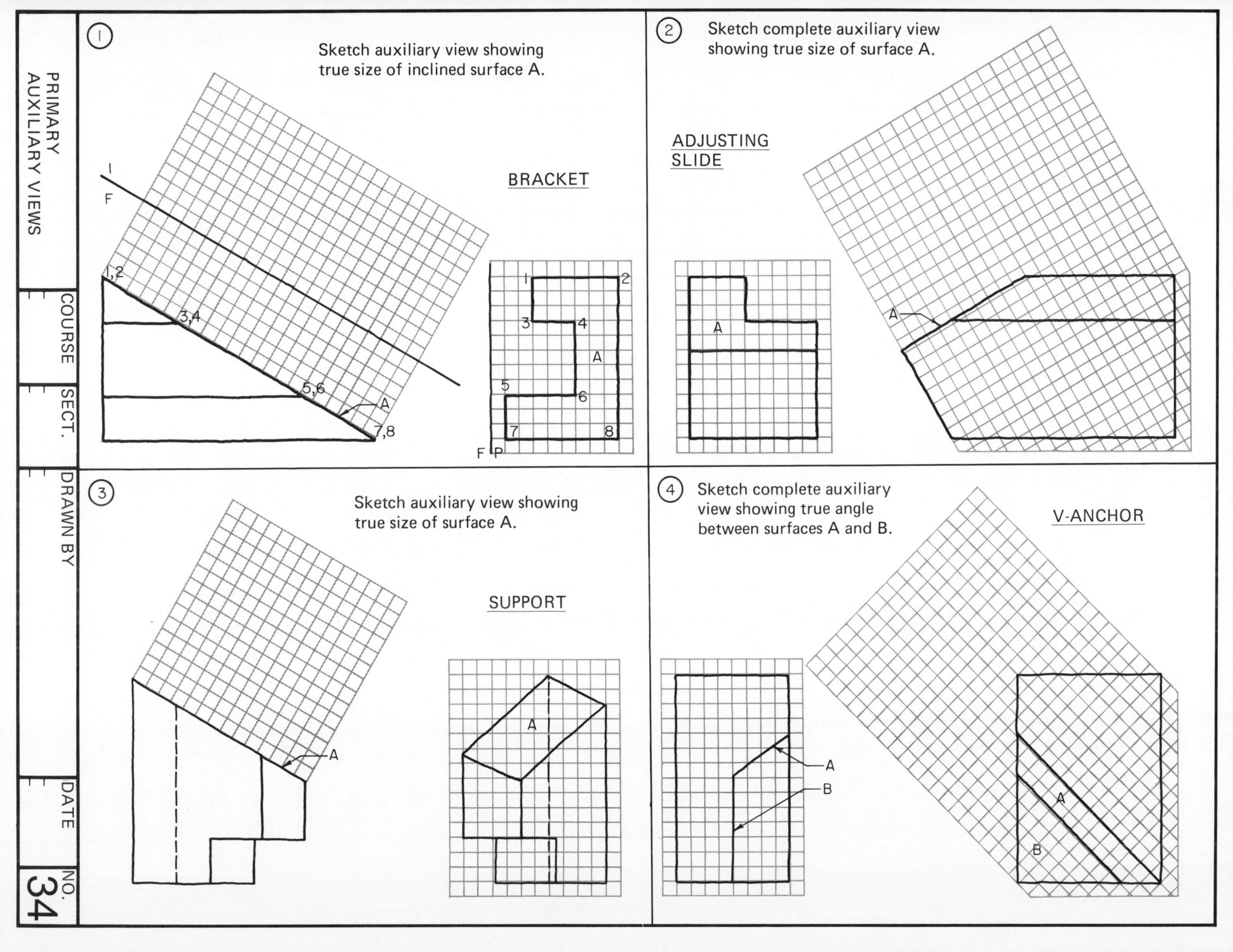
1
Sketch auxiliary view showing
true size of inclined surface A.
BRACKET
1
F
1,2
3,4
5,6
A
7,8
1
2
3
4
A
5
6
7
8
F P
2
Sketch complete auxiliary view
showing true size of surface A.
ADJUSTING
SLIDE
A
A
3
Sketch auxiliary view showing
true size of surface A.
SUPPORT
A
A
4
Sketch complete auxiliary
view showing true angle
between surfaces A and B.
V-ANCHOR
A
B
A
B
PRIMARY
AUXILIARY VIEWS
COURSE
SECT.
DRAWN BY
DATE
NO.
34

① Revolve line 1-2 55° counter-clockwise, about axis 3-4.
Redraw line 1-2 in the front and side views.

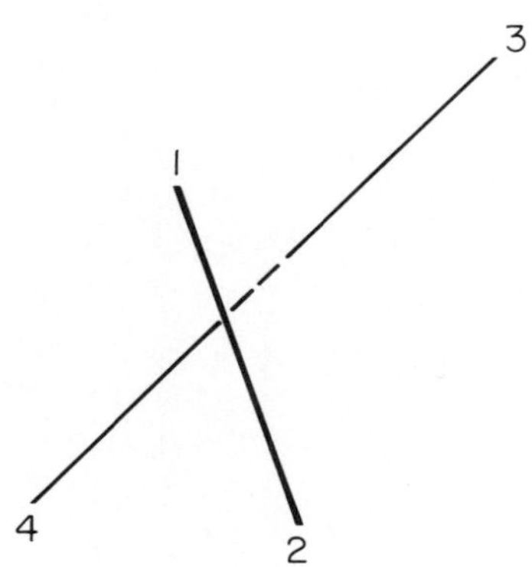

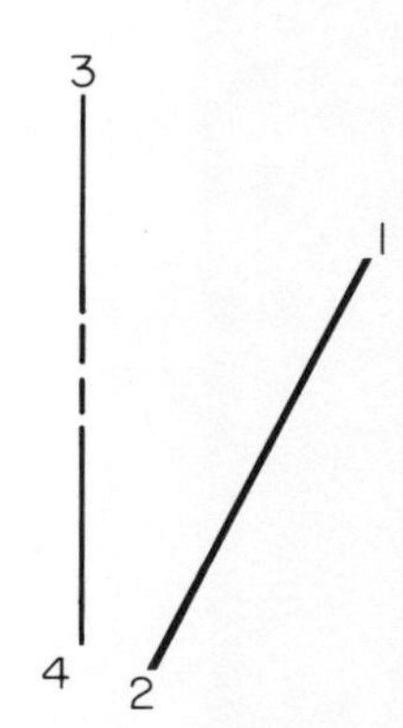

② Revolve the V-Block about point A until the bottom of the milled V slot is vertical. Draw the front top, and right-side views in the revolved position.

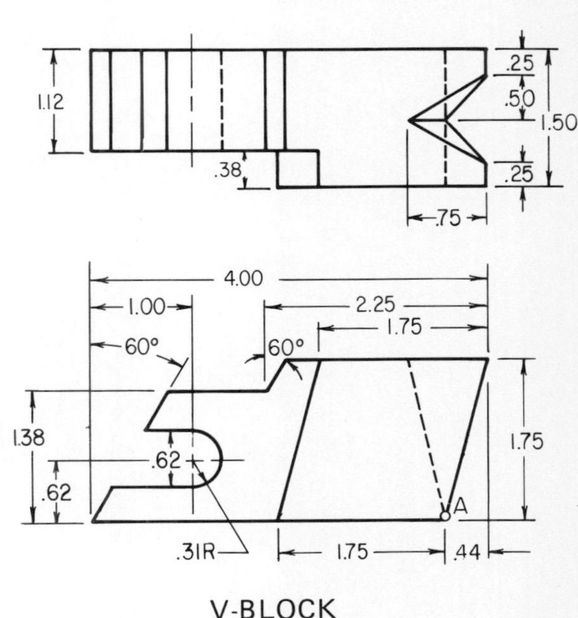

V-BLOCK

○A

REVOLUTION OF LINES AND PLANES	COURSE	SECT.	DRAWN BY	DATE	NO. 41

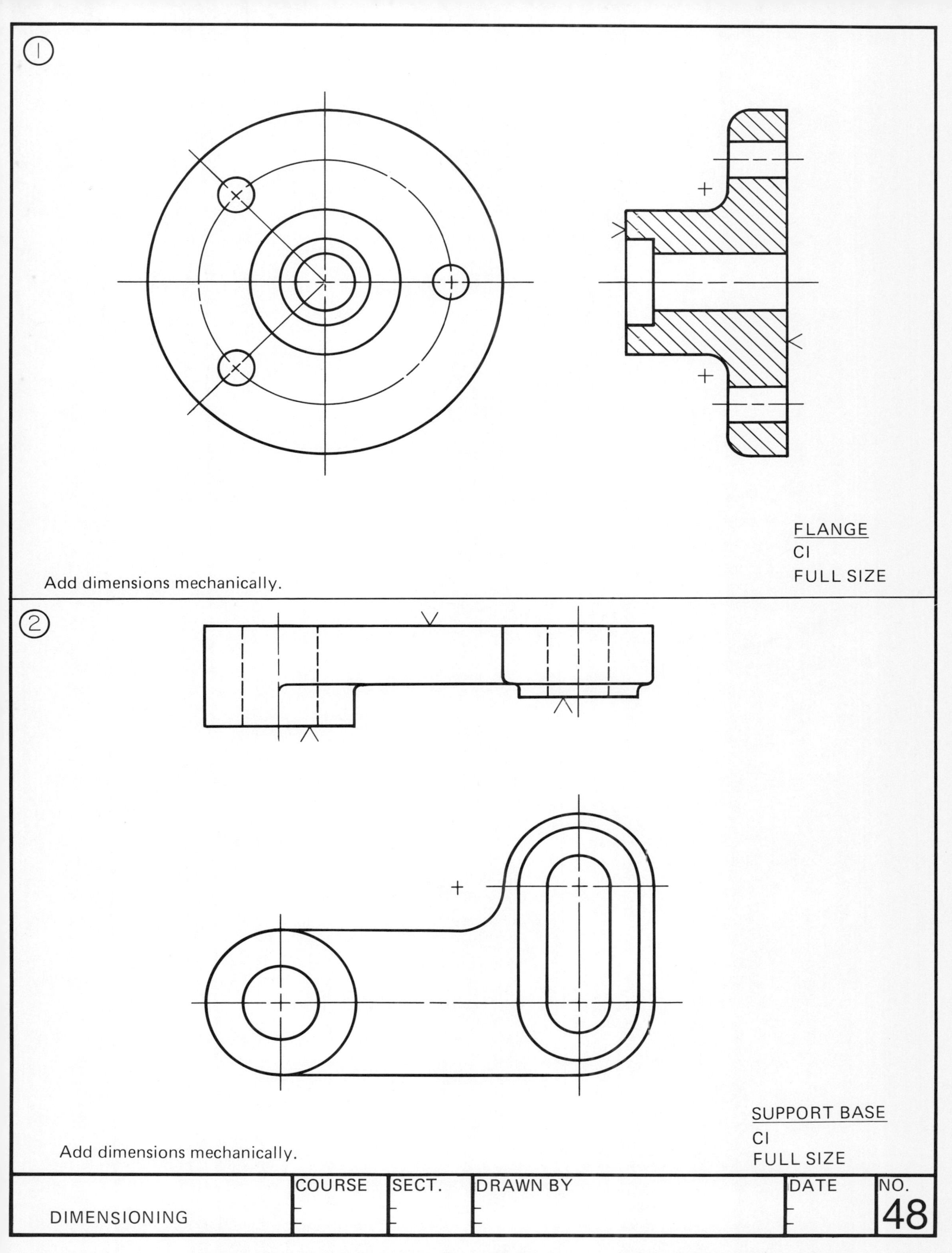
1
FLANGE
CI
FULL SIZE
Add dimensions mechanically.
2
SUPPORT BASE
CI
FULL SIZE
Add dimensions mechanically.
DIMENSIONING
COURSE
SECT.
DRAWN BY
DATE
NO.
48

NEEDLE CORE
BRASS
DOUBLE SIZE

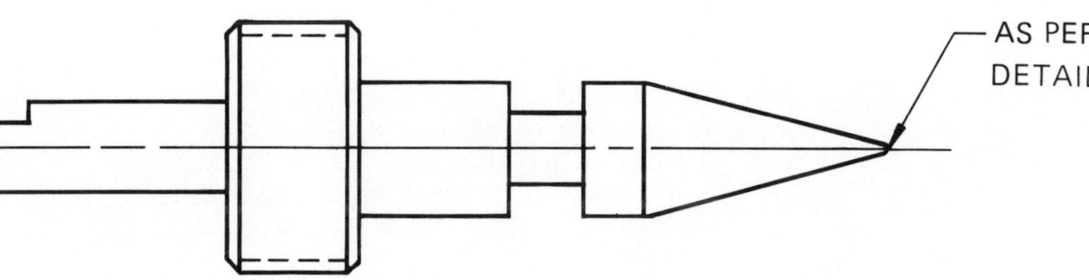

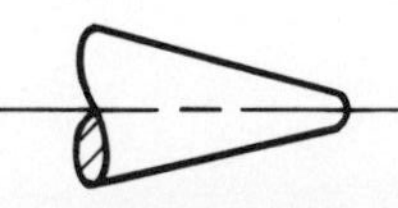

DETAIL A
SCALE: 4:1

METRIC
Add dimensions mechanically.

DIMENSIONING	COURSE	SECT.	DRAWN BY	DATE	NO. 49

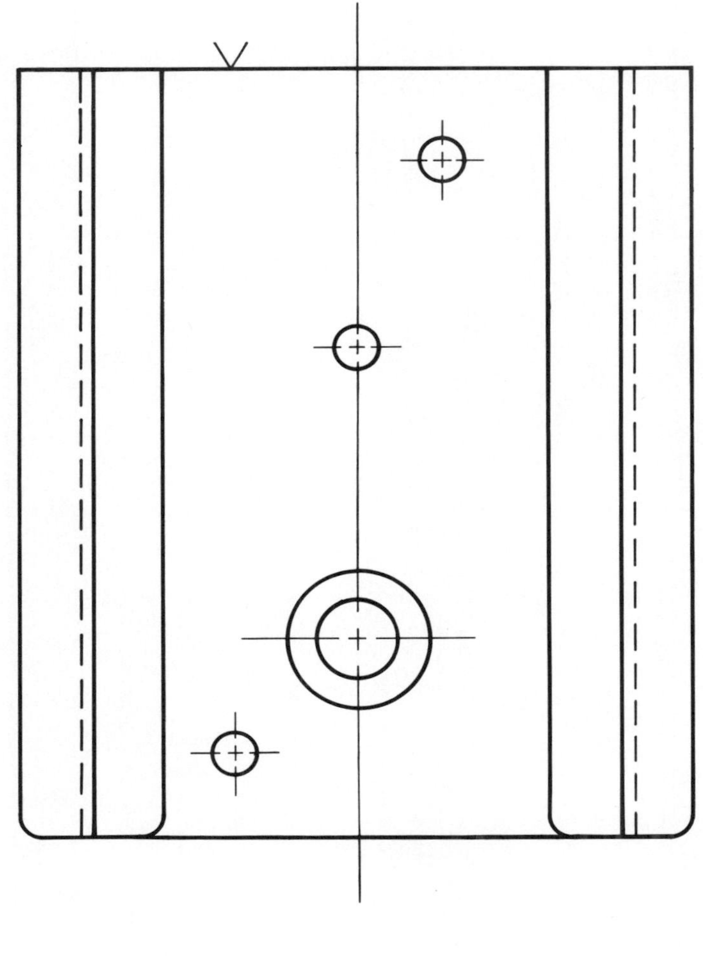

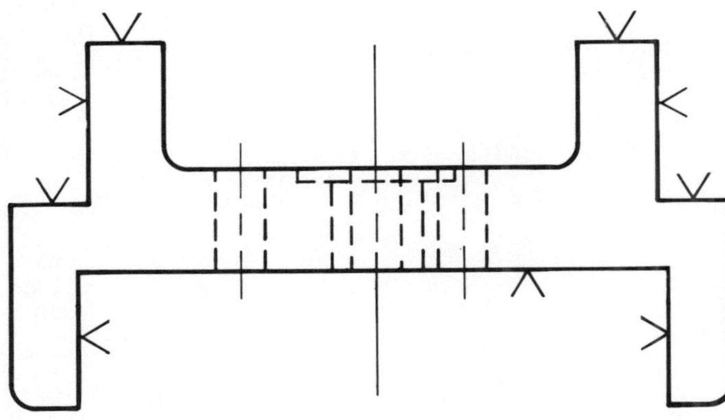

SADDLE BRACKET

CI
FULL SIZE

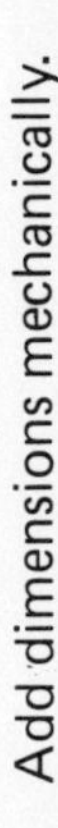

DIMENSIONING	COURSE	SECT.	DRAWN BY	DATE	NO. 51

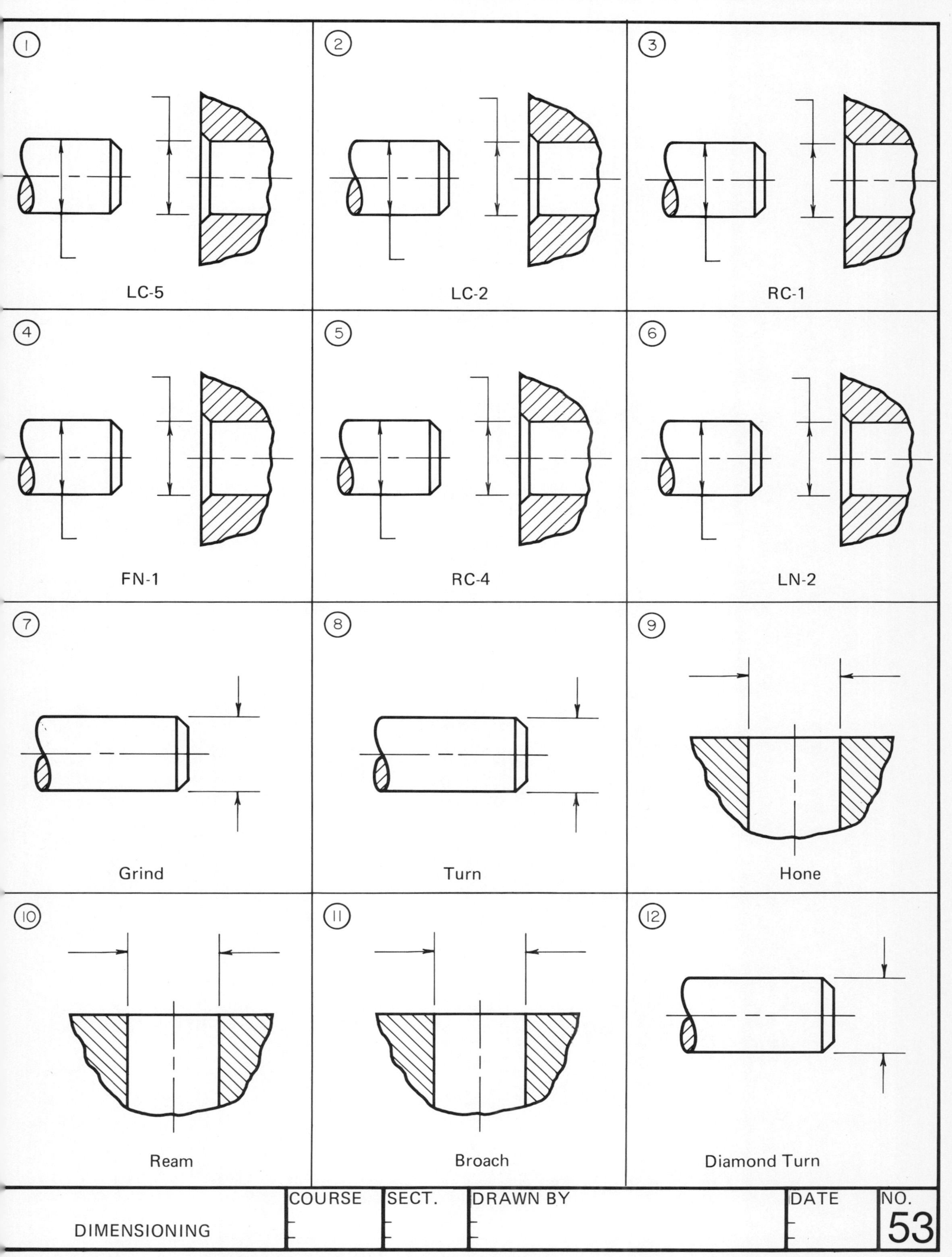
1
LC-5
2
LC-2
3
RC-1
4
FN-1
5
RC-4
6
LN-2
7
Grind
8
Turn
9
Hone
10
Ream
11
Broach
12
Diamond Turn
DIMENSIONING
COURSE
SECT.
DRAWN BY
DATE
NO.
53

Make a detail drawing of the 5″ Caster Fork designed by Walter Dorwin Teague Assoc. Change all fractional dimensions to two place decimals (.XX). The two 5/16″ Dia holes are drilled in line. To clarify the 5/16″ R slot in the lower portion of the fork draw a detail section through the slot at full, 1.5 X or 2 X size. Specify a tolerance of ± .03 on .XX decimals, ± .010 on .XXX decimals and ± 0° 30′ on angular dimensions. Include the following notes on the drawing: 1. Maintain distinctive contour of this design; and 2. All external surfaces must be machine polished only. The fork is cast from 40 E A1. (J&J, I)

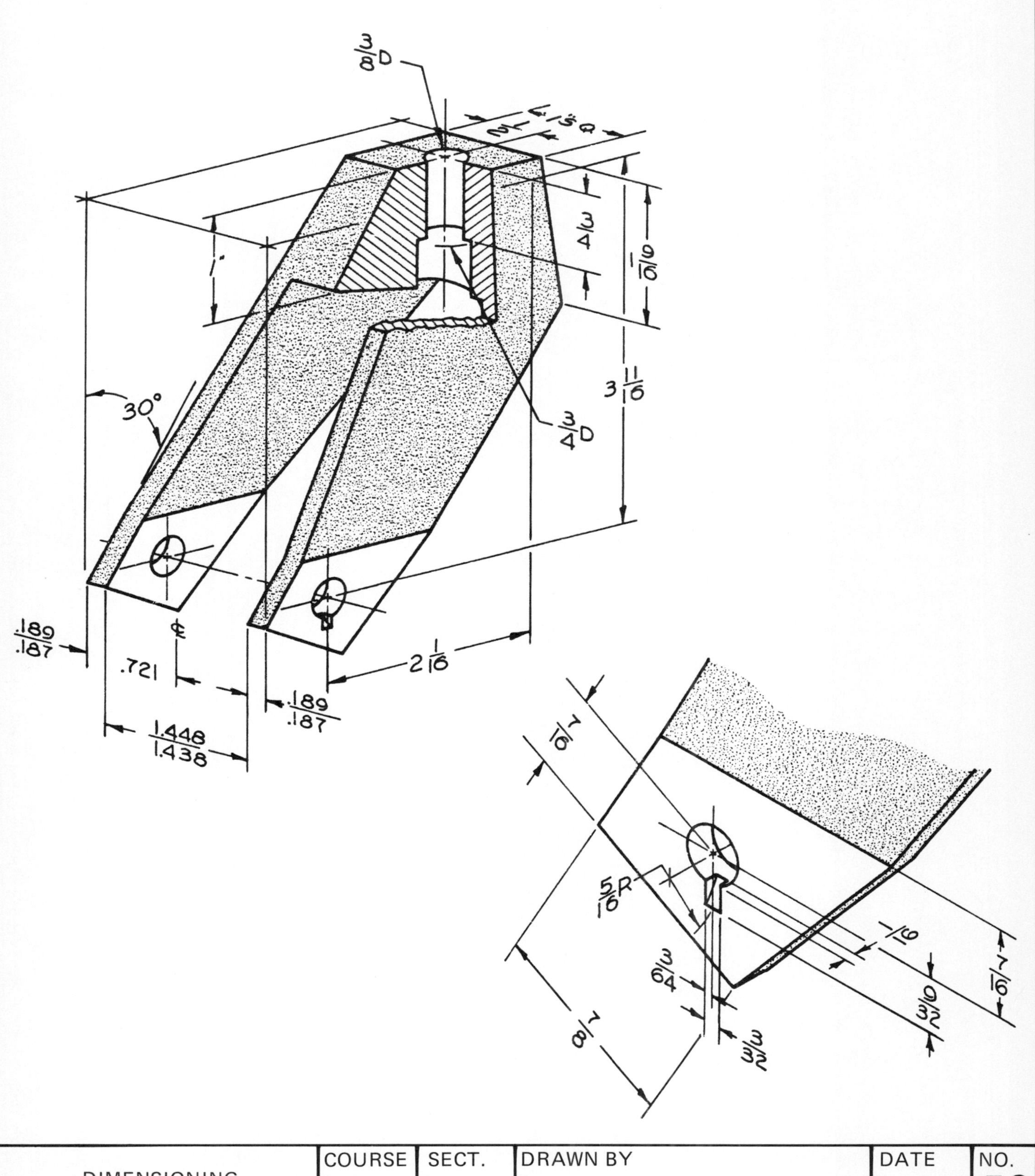

DIMENSIONING	COURSE	SECT.	DRAWN BY	DATE	NO. 56

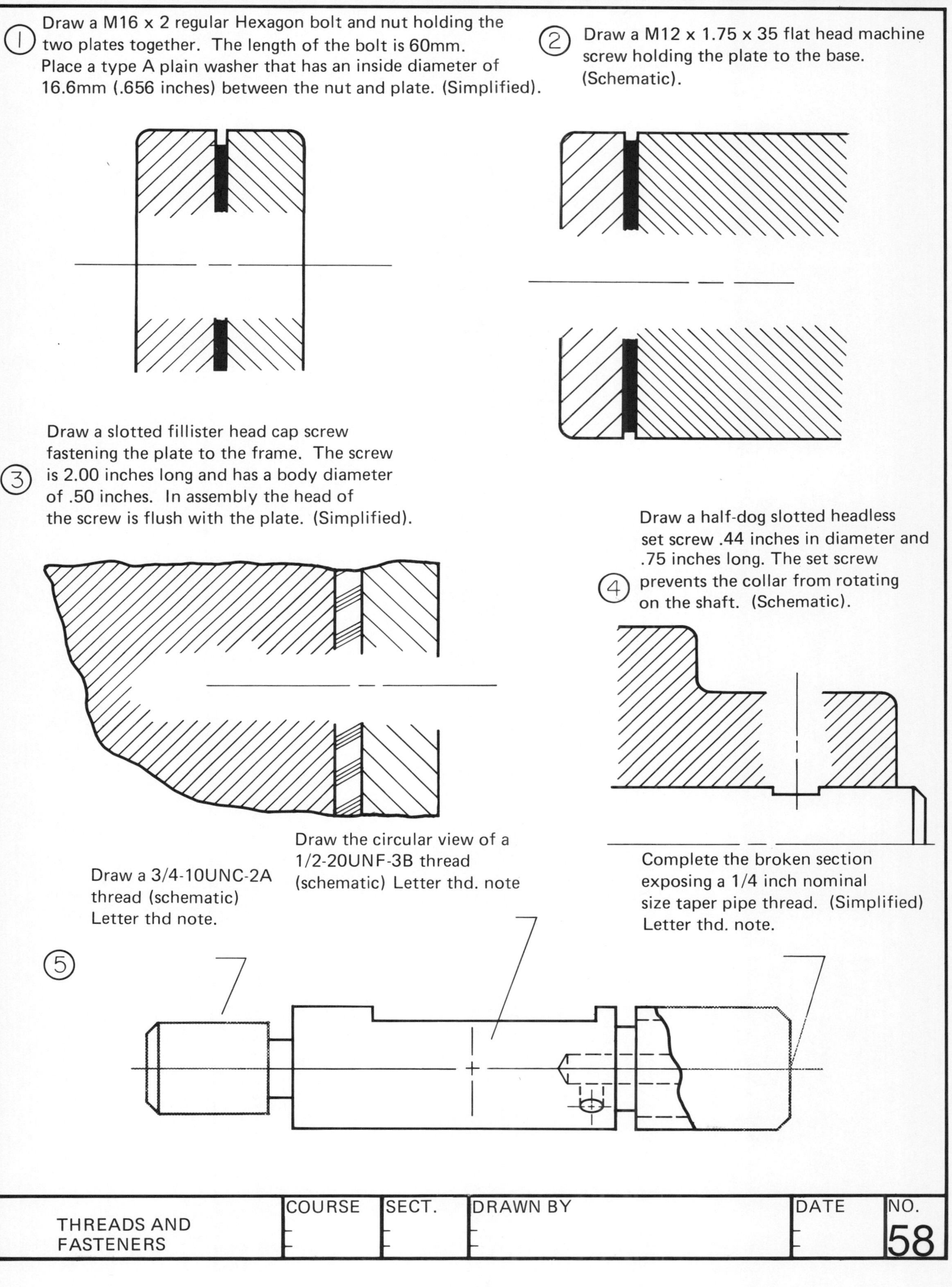

1
Draw a M16 x 2 regular Hexagon bolt and nut holding the two plates together. The length of the bolt is 60mm. Place a type A plain washer that has an inside diameter of 16.6mm (.656 inches) between the nut and plate. (Simplified).
2
Draw a M12 x 1.75 x 35 flat head machine screw holding the plate to the base. (Schematic).
3
Draw a slotted fillister head cap screw fastening the plate to the frame. The screw is 2.00 inches long and has a body diameter of .50 inches. In assembly the head of the screw is flush with the plate. (Simplified).
4
Draw a half-dog slotted headless set screw .44 inches in diameter and .75 inches long. The set screw prevents the collar from rotating on the shaft. (Schematic).
5
Draw a 3/4-10UNC-2A thread (schematic) Letter thd note.
Draw the circular view of a 1/2-20UNF-3B thread (schematic) Letter thd. note
Complete the broken section exposing a 1/4 inch nominal size taper pipe thread. (Simplified) Letter thd. note.
THREADS AND FASTENERS
COURSE
SECT.
DRAWN BY
DATE
NO.
58

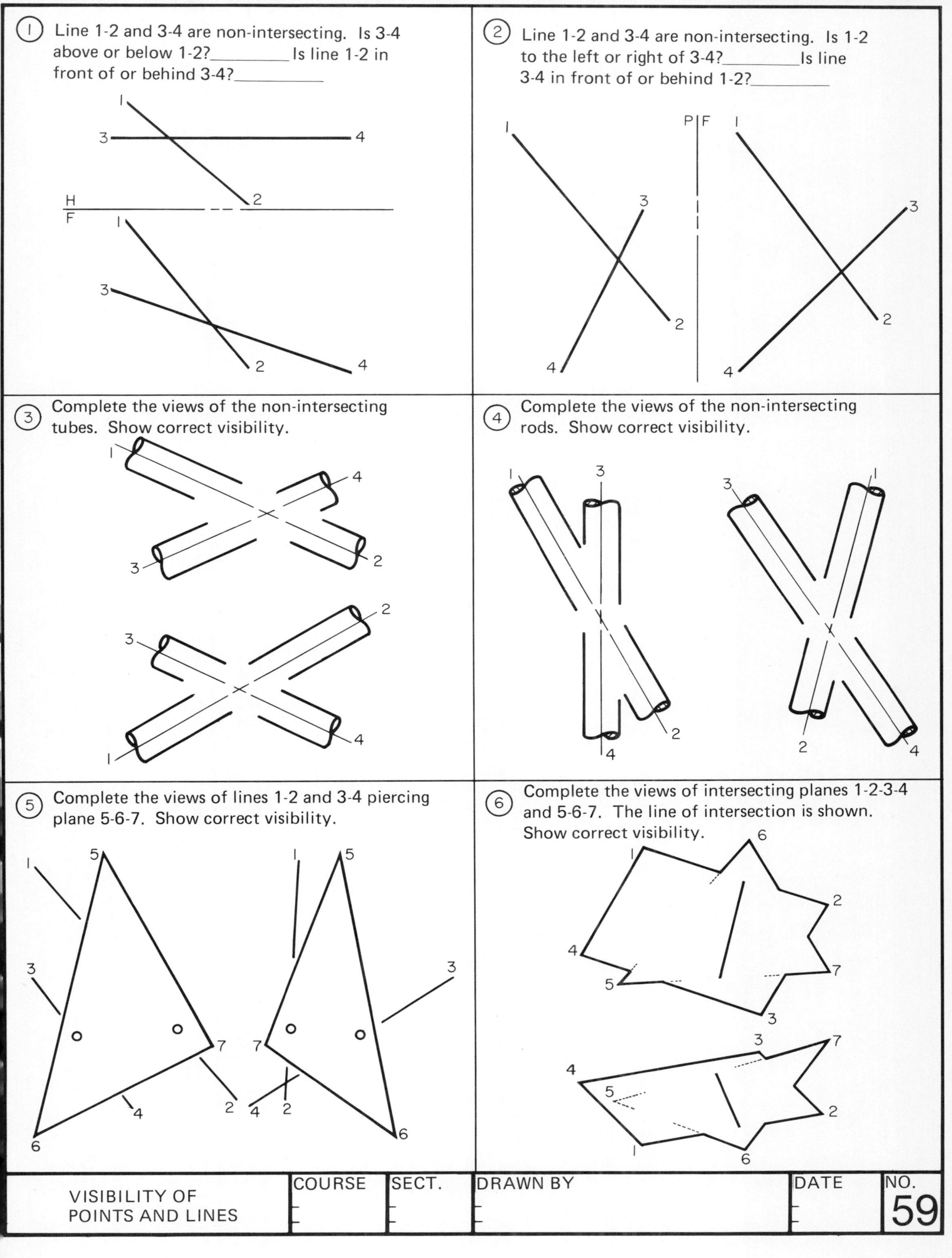

VISIBILITY OF POINTS AND LINES	COURSE	SECT.	DRAWN BY	DATE	NO. 59

AUXILIARY VIEW TEACHING AID

An auxiliary view is an extra view added to the principle views to obtain the true length of a line. The auxiliary plane is placed parallel to one of the principle views of the line. In the drawing below, the auxiliary reference plane (fold line) is located parallel to the top view of line AB. When an auxiliary reference line is located parallel to any line, *that line* will appear fundamental (true length) in *that* auxiliary view. The auxiliary view below shows line AB in true length as well as the true angle (slope) it makes with the horizontal plane. The bearing appears only in the top view. To better understand the relationship between principle views and auxiliary views fold the sheet as indicated below.

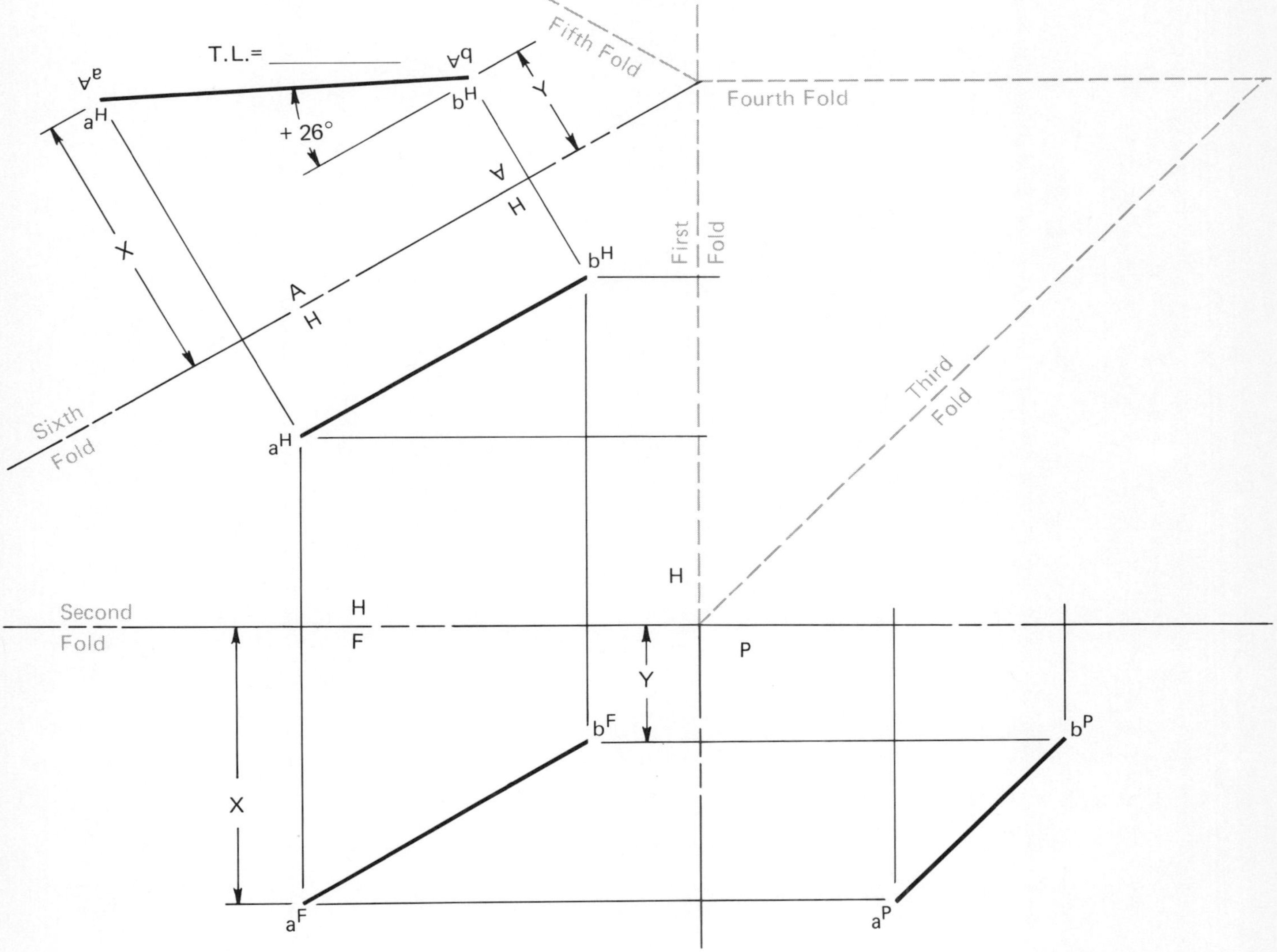

T.L.= ____________

+ 26°

Fifth Fold

Fourth Fold

First Fold

Third Fold

Sixth Fold

Second Fold

a^A b^A a^H b^H a^F b^F a^P b^P

A/H   H/F   H/P   X   Y

FOLDING DIRECTIONS FOR TEACHING AID

1. Fold sheet back and forth along each fold line to make sheet flexible (first fold is along entire length of sheet)
2. Fold sheet lengthwise along first fold.
3. Push third fold down and in to form corner between Frontal and Horizontal planes.
4. Fold fourth and fifth folds under leaving triangular shaped top. (Fig. 4 and 5)
5. Fold along sixth fold pushing auxillary plane down perpendicular to Horizontal plane.
6. Figures 6 and 7 show completed views of folded sheet.

② 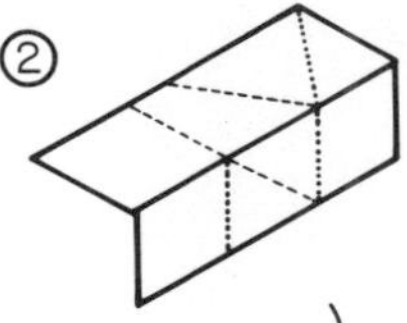

③ 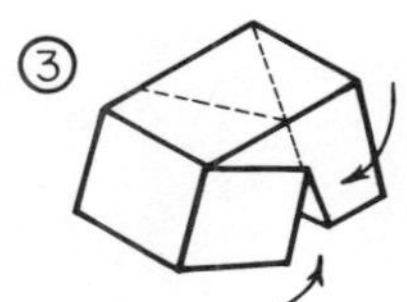

④ 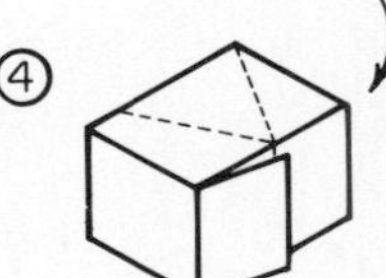

⑤ 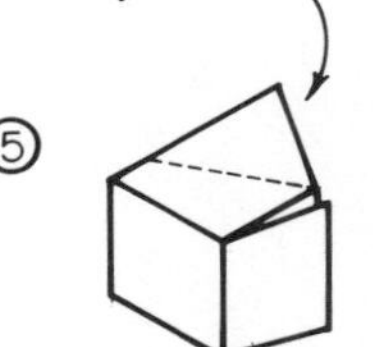

⑥ 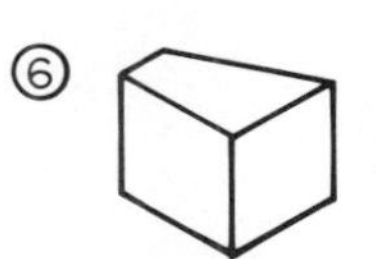

Completed sheet

⑦ 
Rear view.

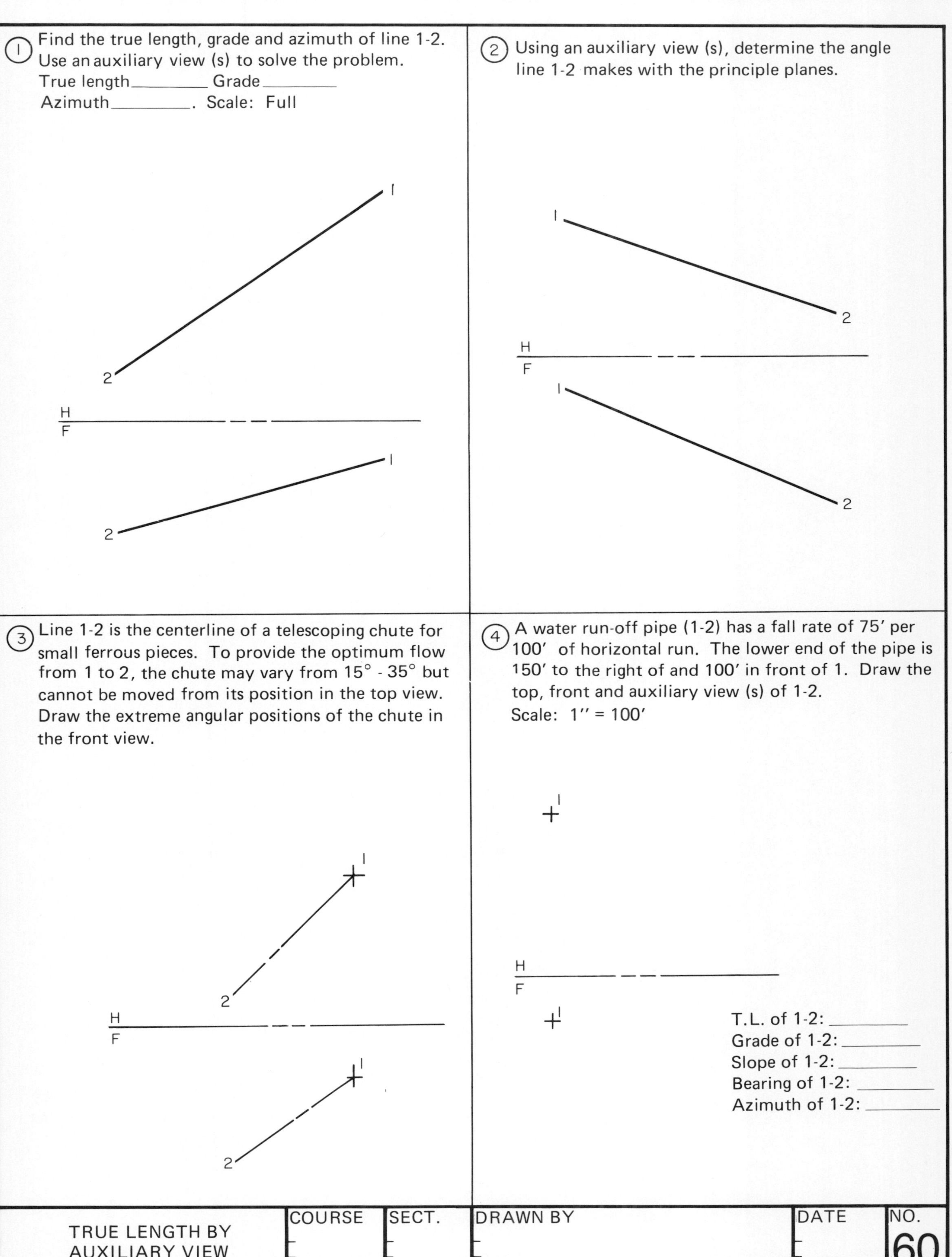

1 Find the true length, grade and azimuth of line 1-2. Use an auxiliary view (s) to solve the problem.
True length________ Grade________
Azimuth________. Scale: Full
1
2
H
F
1
2
2 Using an auxiliary view (s), determine the angle line 1-2 makes with the principle planes.
1
2
H
F
1
2
3 Line 1-2 is the centerline of a telescoping chute for small ferrous pieces. To provide the optimum flow from 1 to 2, the chute may vary from 15° - 35° but cannot be moved from its position in the top view. Draw the extreme angular positions of the chute in the front view.
1
2
H
F
1
2
4 A water run-off pipe (1-2) has a fall rate of 75′ per 100′ of horizontal run. The lower end of the pipe is 150′ to the right of and 100′ in front of 1. Draw the top, front and auxiliary view (s) of 1-2.
Scale: 1″ = 100′
1
H
F
1
T.L. of 1-2: ________
Grade of 1-2: ________
Slope of 1-2: ________
Bearing of 1-2: ________
Azimuth of 1-2: ________
TRUE LENGTH BY AUXILIARY VIEW
COURSE
SECT.
DRAWN BY
DATE
NO.
60

(1) Find the true length, bearing and slope of line 1-2 using revolution. Scale: Full

(2) Find the angle formed by line 1-2 and the principle planes.

(3) A liquid supply line 1-2 may have a minimum and maximum slope of −35° and −55° respectively. The position of 1-2 in the top view cannot be changed. Draw the front and top views showing the alternate extreme positions of the supply line.

(4) A line has a bearing of S 57° E and a slope of −45° and a true length of 200.00 feet. Draw the top and front views of the line. (Scale: 1″=100′)

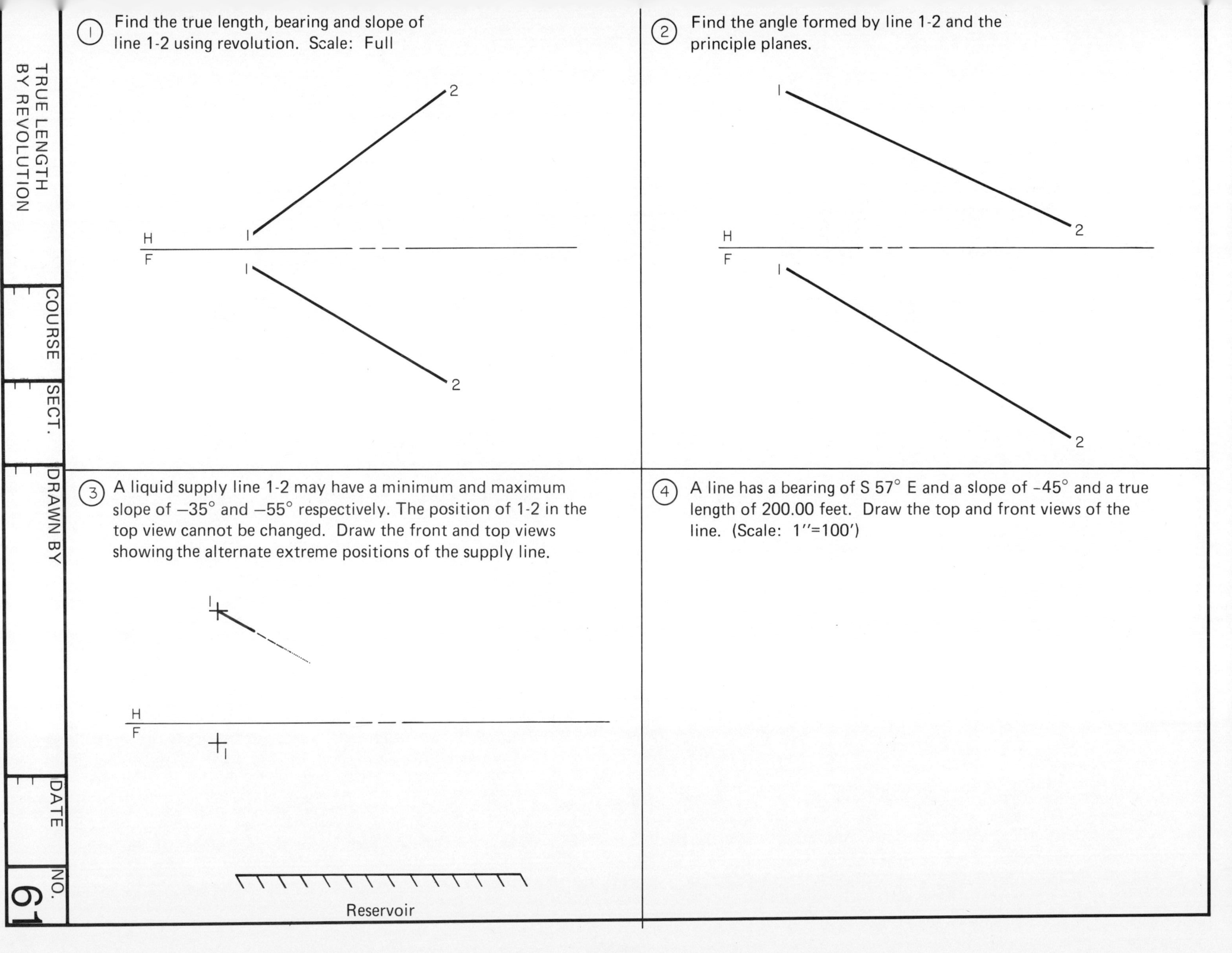

TRUE LENGTH BY REVOLUTION	COURSE	SECT.	DRAWN BY	DATE	NO. 61

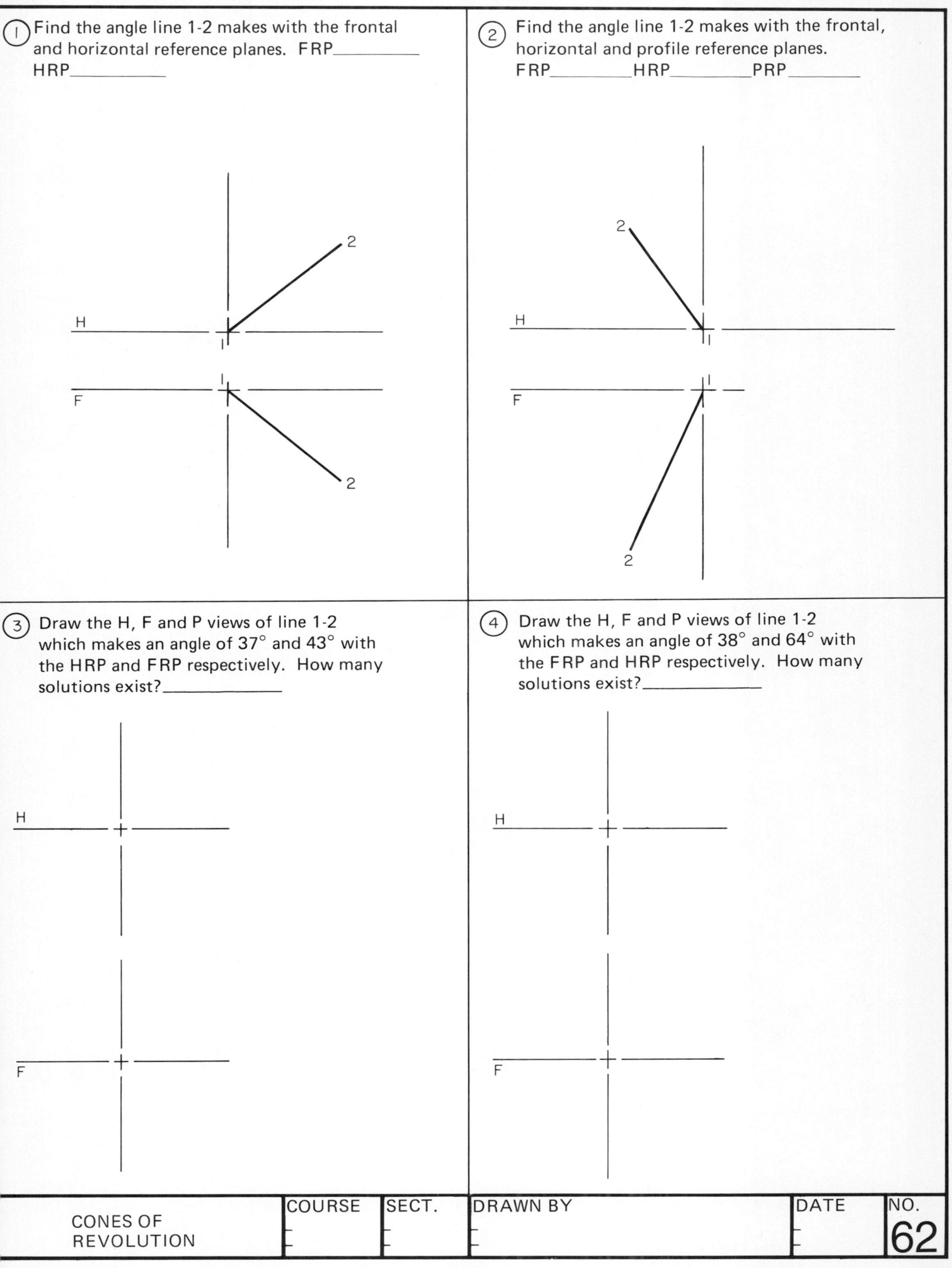
1 Find the angle line 1-2 makes with the frontal and horizontal reference planes. FRP________ HRP________
2
H
1
1
F
2
2 Find the angle line 1-2 makes with the frontal, horizontal and profile reference planes.
FRP________HRP________PRP________
2
H
1
1
F
2
3 Draw the H, F and P views of line 1-2 which makes an angle of 37° and 43° with the HRP and FRP respectively. How many solutions exist?________
H
F
4 Draw the H, F and P views of line 1-2 which makes an angle of 38° and 64° with the FRP and HRP respectively. How many solutions exist?________
H
F
CONES OF REVOLUTION
COURSE
SECT.
DRAWN BY
DATE
NO.
62

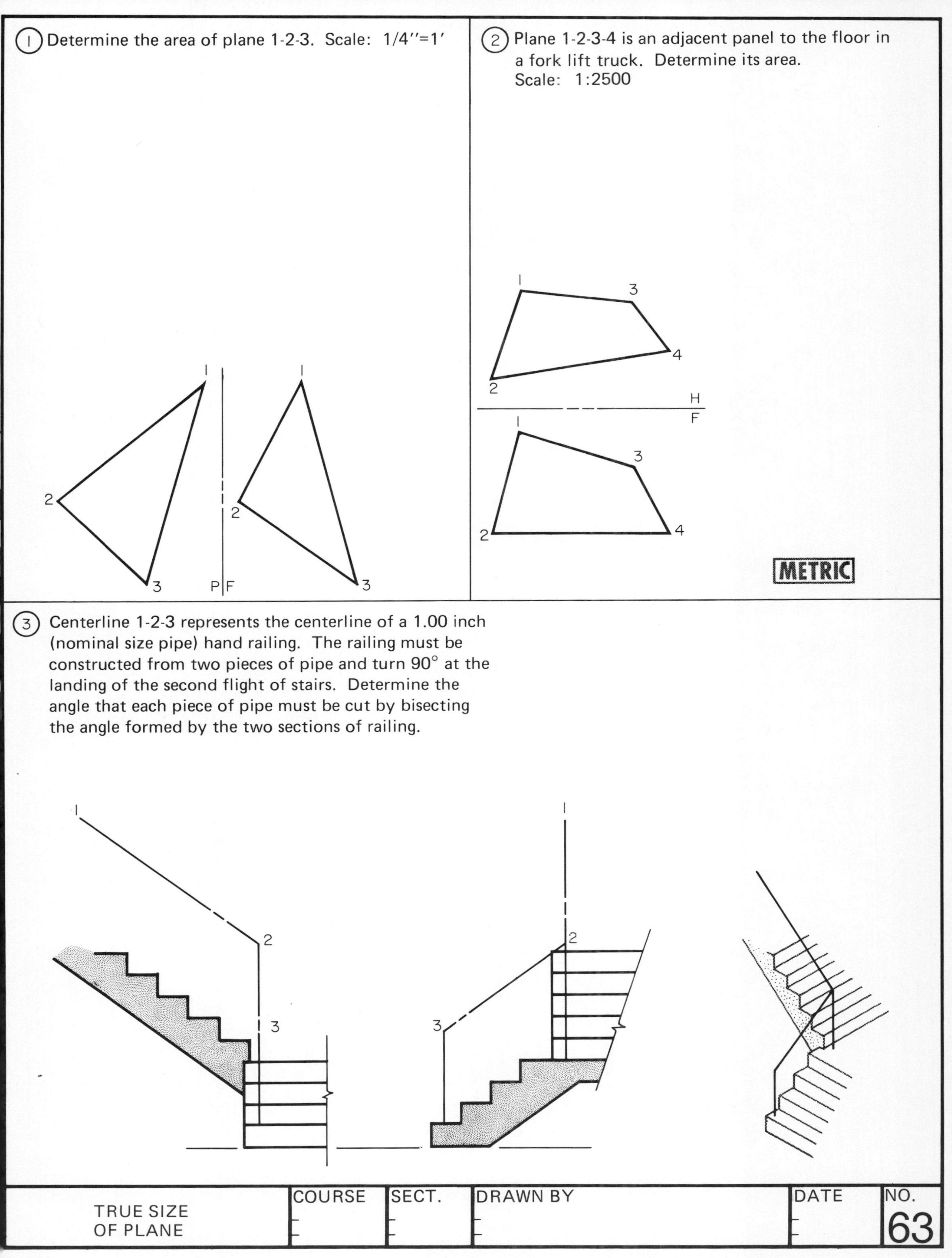

① Determine the area of plane 1-2-3. Scale: 1/4″=1′
1
2
3
P
F
1
2
3
② Plane 1-2-3-4 is an adjacent panel to the floor in a fork lift truck. Determine its area.
Scale: 1:2500
1
3
4
2
H
F
1
3
2
4
METRIC
③ Centerline 1-2-3 represents the centerline of a 1.00 inch (nominal size pipe) hand railing. The railing must be constructed from two pieces of pipe and turn 90° at the landing of the second flight of stairs. Determine the angle that each piece of pipe must be cut by bisecting the angle formed by the two sections of railing.
1
2
3
1
2
3
TRUE SIZE
OF PLANE
COURSE
SECT.
DRAWN BY
DATE
NO.
63

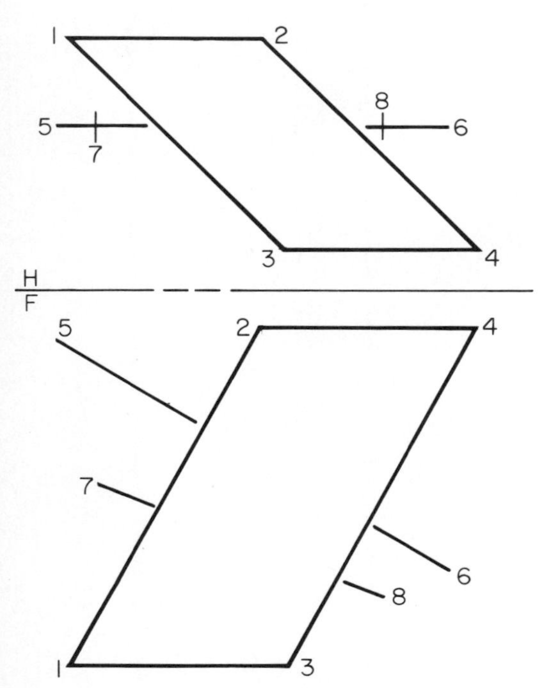

Plane 1-2-3-4 is the windshield of a model crane. The boom cable (5-6) controlling the bucket passes through the windshield. The extreme position of the cable is shown by line 7-8. Find, by using auxiliary(s) views, the piercing points of the extreme positions of the cable.

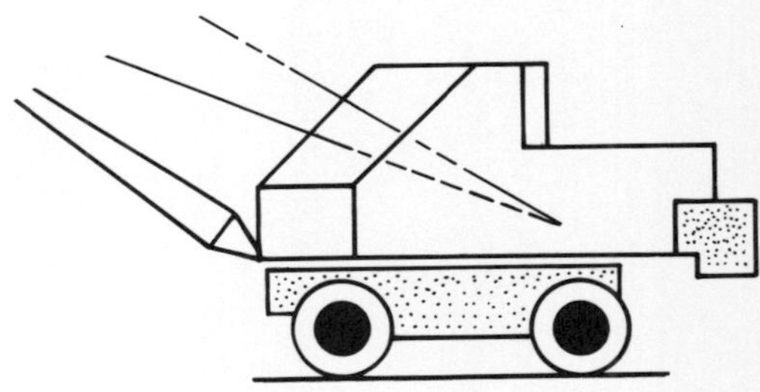

② An experimental tracer bullet (line 5-6) is fired at point A on oblique panel 1-2-3-4. Determine by using auxiliary view(s) if the bullet missed point A.

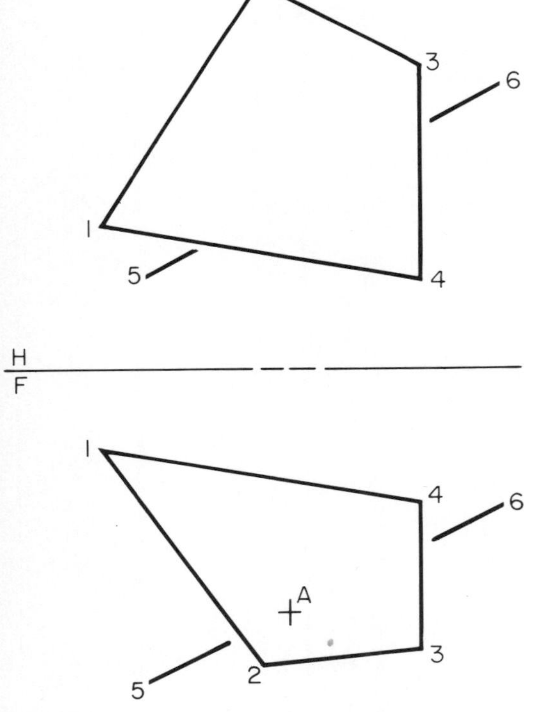

PIERCING POINTS EDGE-VIEW METHOD	COURSE	SECT.	DRAWN BY	DATE	NO. 64

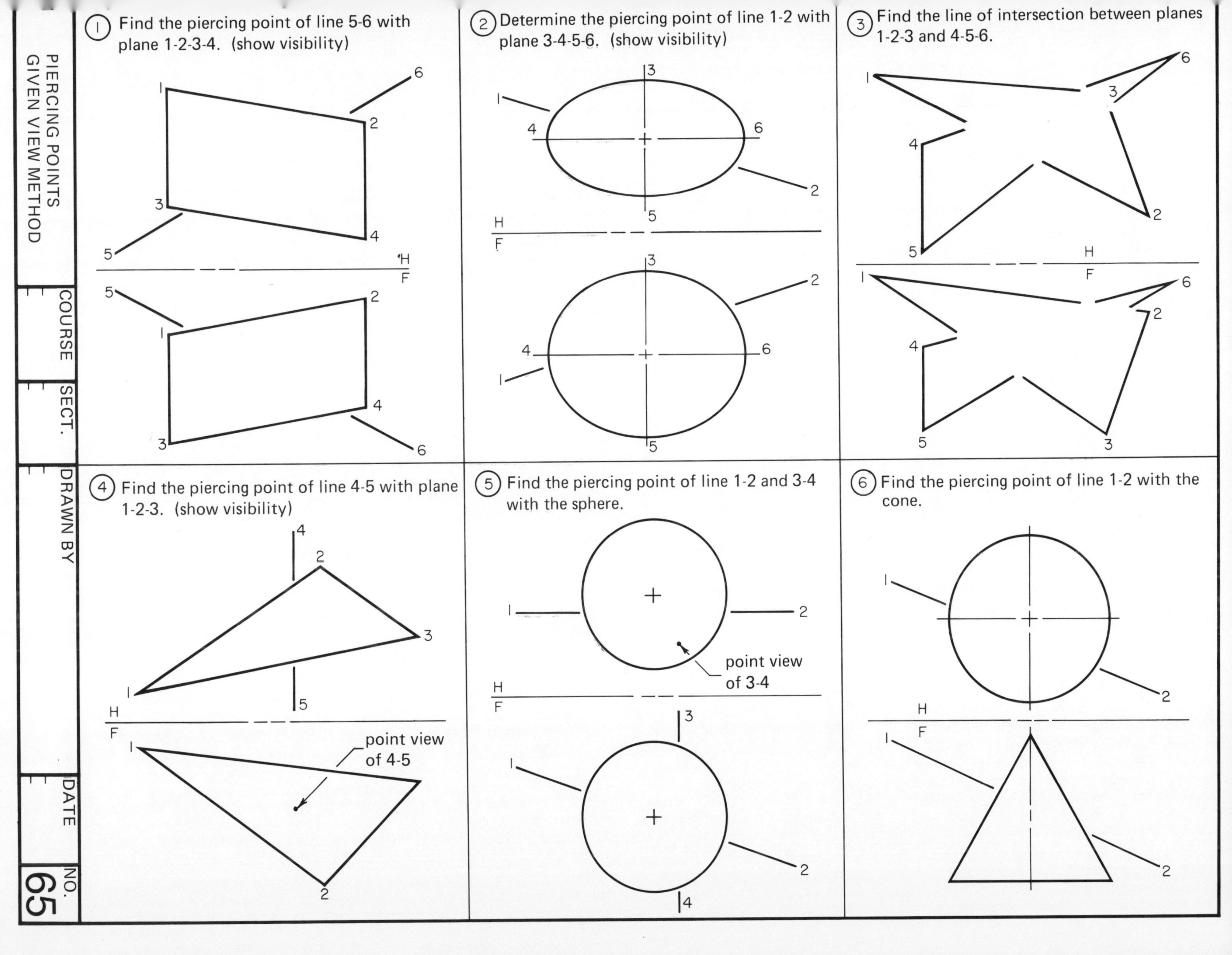
(1) Find the piercing point of line 5-6 with plane 1-2-3-4. (show visibility)
6
1
2
3
4
5
H
F
5
2
1
4
3
6
(2) Determine the piercing point of line 1-2 with plane 3-4-5-6. (show visibility)
3
1
4
6
2
5
H
F
3
2
4
6
1
5
(3) Find the line of intersection between planes 1-2-3 and 4-5-6.
6
1
3
4
2
5
H
F
1
6
2
4
5
3
(4) Find the piercing point of line 4-5 with plane 1-2-3. (show visibility)
4
2
3
1
5
H
F
1
point view of 4-5
2
(5) Find the piercing point of line 1-2 and 3-4 with the sphere.
1
2
point view of 3-4
H
F
3
1
2
4
(6) Find the piercing point of line 1-2 with the cone.
1
2
H
F
1
2
PIERCING POINTS
GIVEN VIEW METHOD
COURSE
SECT.
DRAWN BY
DATE
NO.
65

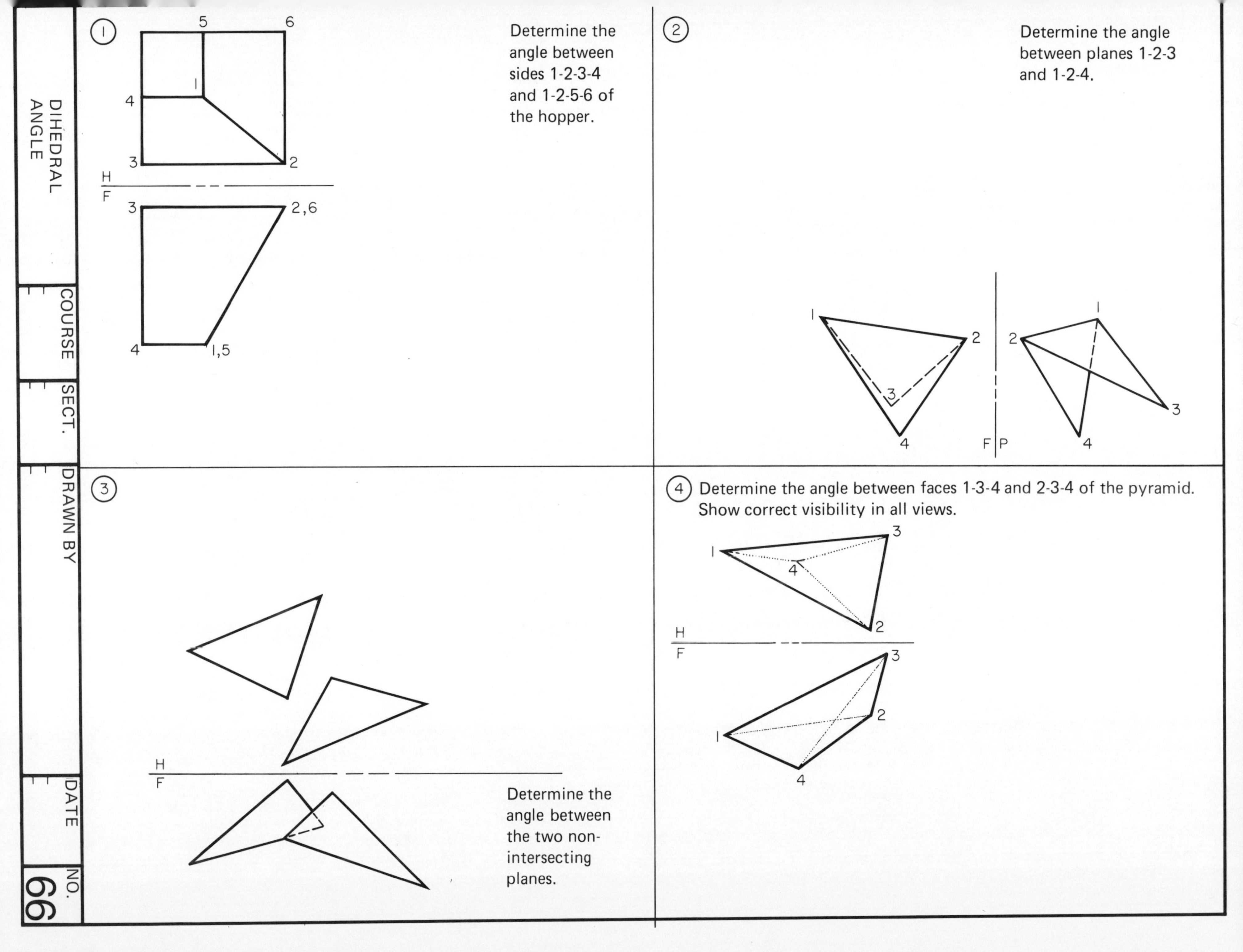
DIHEDRAL ANGLE
COURSE
SECT.
DRAWN BY
DATE
NO.
66
1
Determine the angle between sides 1-2-3-4 and 1-2-5-6 of the hopper.
H
F
2
Determine the angle between planes 1-2-3 and 1-2-4.
F P
3
Determine the angle between the two non-intersecting planes.
H
F
4
Determine the angle between faces 1-3-4 and 2-3-4 of the pyramid. Show correct visibility in all views.
H
F

In order to design a line bracket, find the angle formed by line 1-2 and plane 3-4-5-6. ANS.________

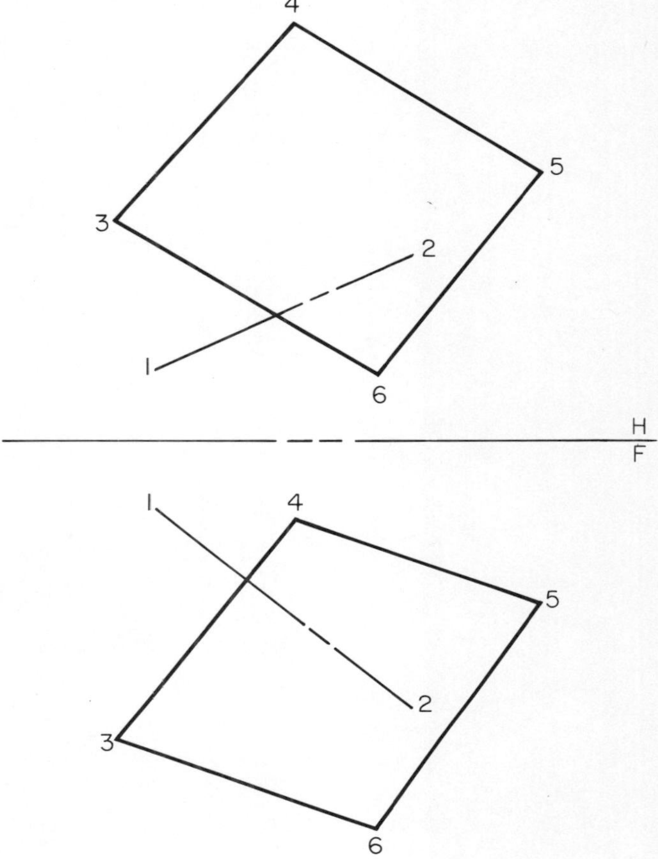

ANGLE BETWEEN LINE AND PLANE	COURSE	SECT.	DRAWN BY	DATE	NO. 67

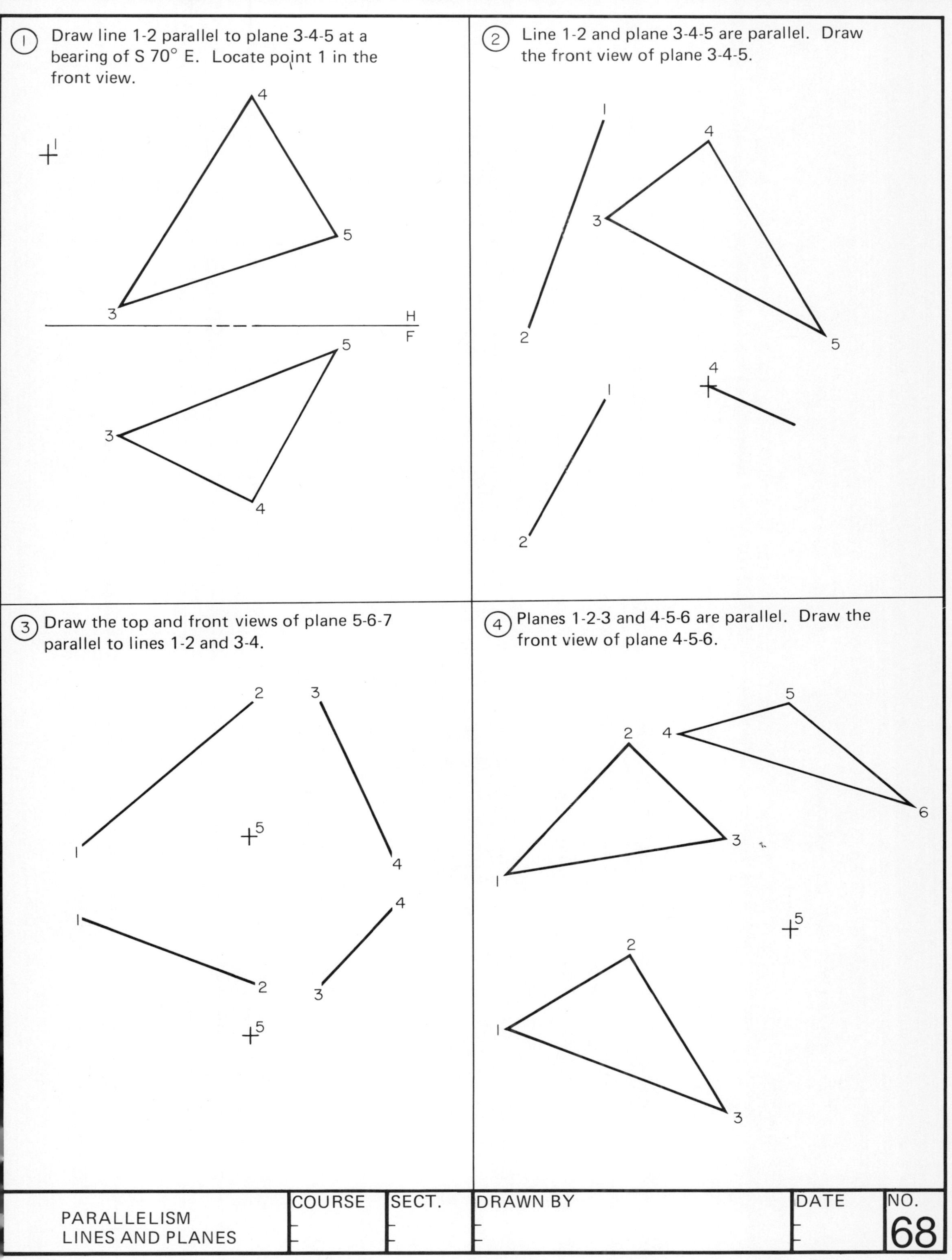

1 Draw line 1-2 parallel to plane 3-4-5 at a bearing of S 70° E. Locate point 1 in the front view.
1
4
5
3
H
F
5
3
4
2 Line 1-2 and plane 3-4-5 are parallel. Draw the front view of plane 3-4-5.
1
4
3
2
5
4
1
2
3 Draw the top and front views of plane 5-6-7 parallel to lines 1-2 and 3-4.
2
3
5
1
4
4
1
2
3
5
4 Planes 1-2-3 and 4-5-6 are parallel. Draw the front view of plane 4-5-6.
5
2
4
6
3
1
5
2
1
3
PARALLELISM
LINES AND PLANES
COURSE
SECT.
DRAWN BY
DATE
NO.
68

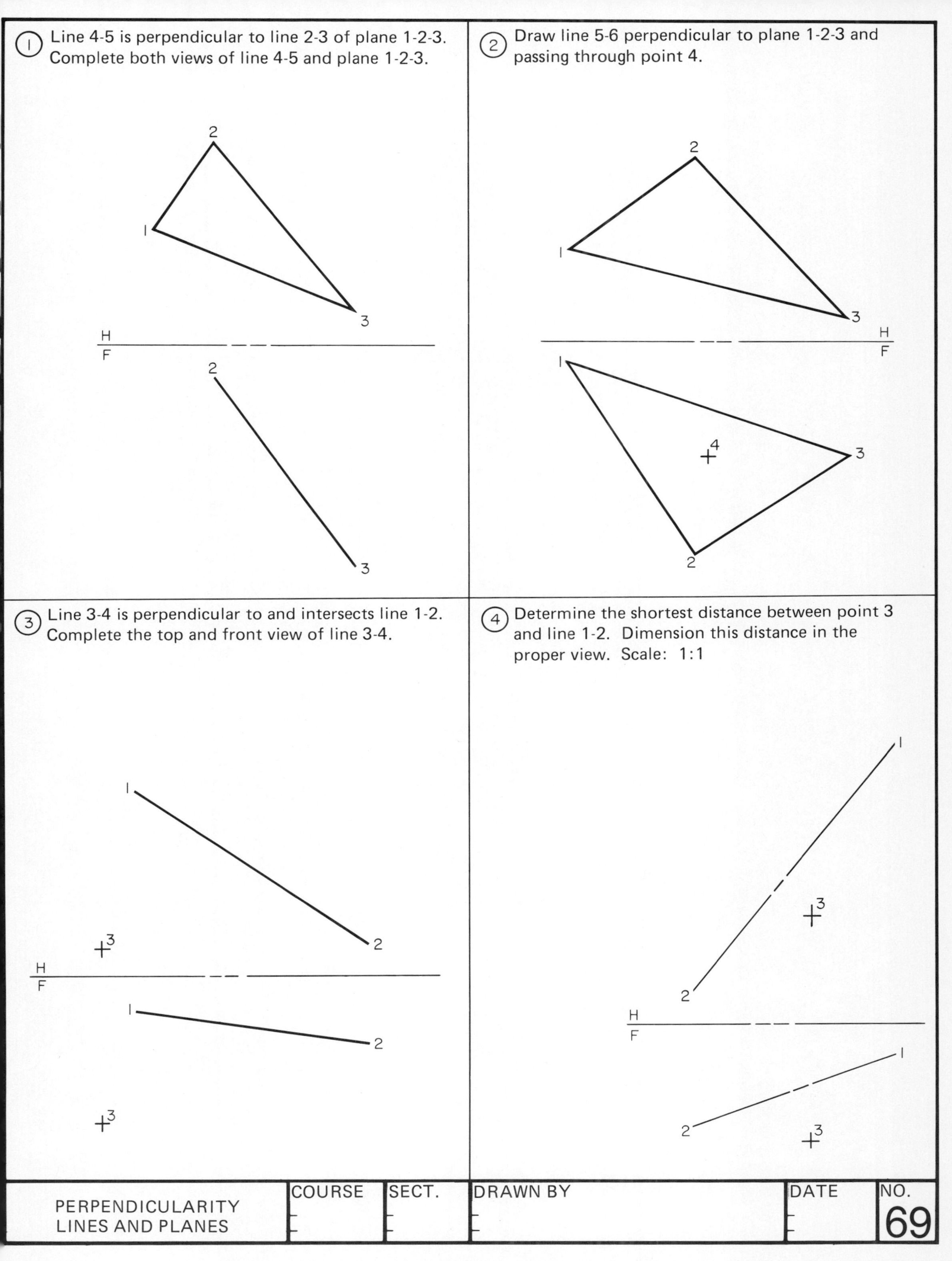

1 Line 4-5 is perpendicular to line 2-3 of plane 1-2-3.
Complete both views of line 4-5 and plane 1-2-3.
2
1
3
H
F
2
3
2 Draw line 5-6 perpendicular to plane 1-2-3 and
passing through point 4.
2
1
3
H
F
1
4
3
2
3 Line 3-4 is perpendicular to and intersects line 1-2.
Complete the top and front view of line 3-4.
1
3
2
H
F
1
2
3
4 Determine the shortest distance between point 3
and line 1-2. Dimension this distance in the
proper view. Scale: 1:1
1
3
2
H
F
1
2
3
PERPENDICULARITY
LINES AND PLANES
COURSE
SECT.
DRAWN BY
DATE
NO.
69

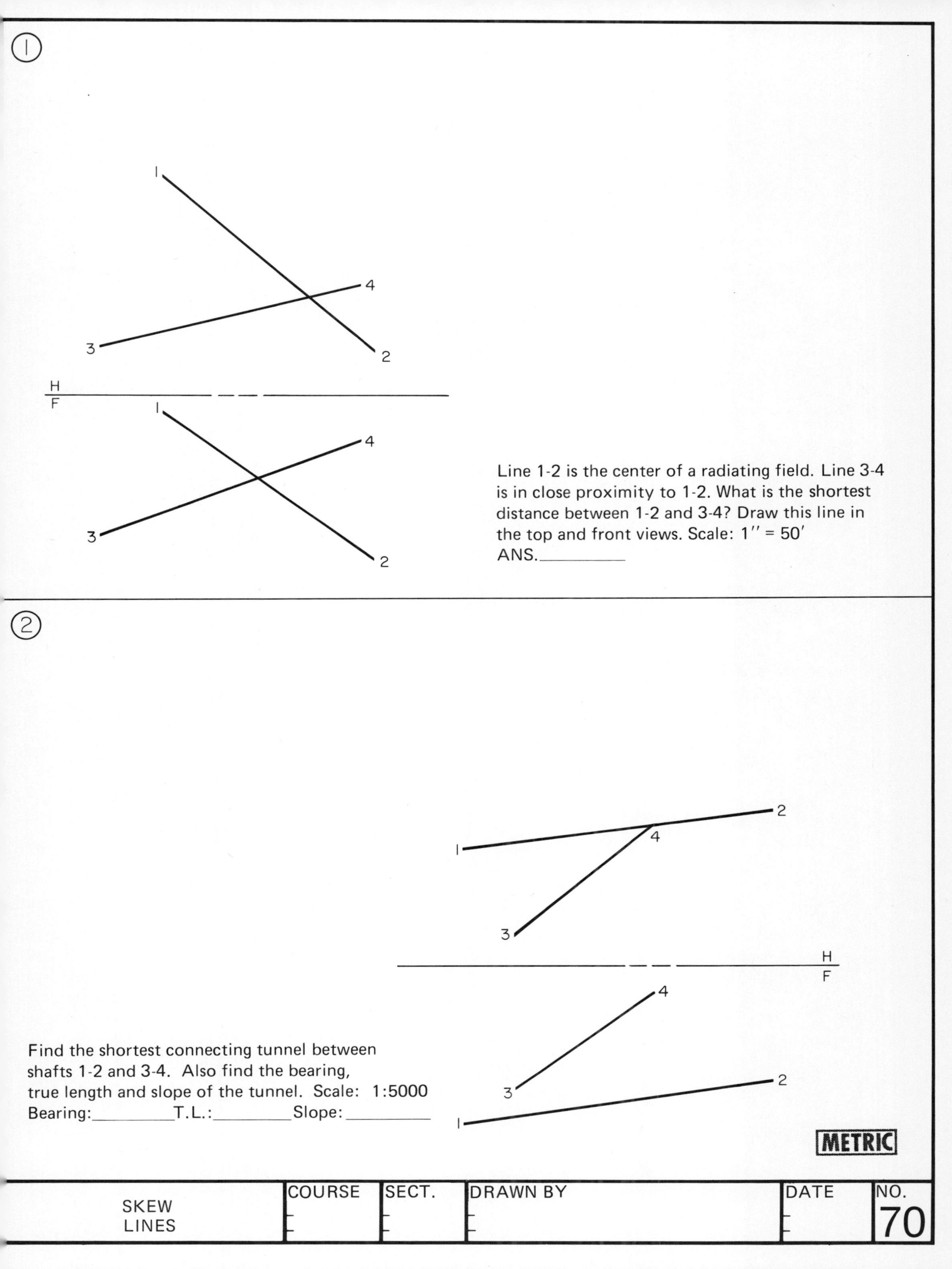

(1)

Line 1-2 is the center of a radiating field. Line 3-4 is in close proximity to 1-2. What is the shortest distance between 1-2 and 3-4? Draw this line in the top and front views. Scale: 1″ = 50′
ANS.________

(2)

Find the shortest connecting tunnel between shafts 1-2 and 3-4. Also find the bearing, true length and slope of the tunnel. Scale: 1:5000
Bearing:________ T.L.:________ Slope:________

SKEW LINES	COURSE	SECT.	DRAWN BY	DATE	NO. 70

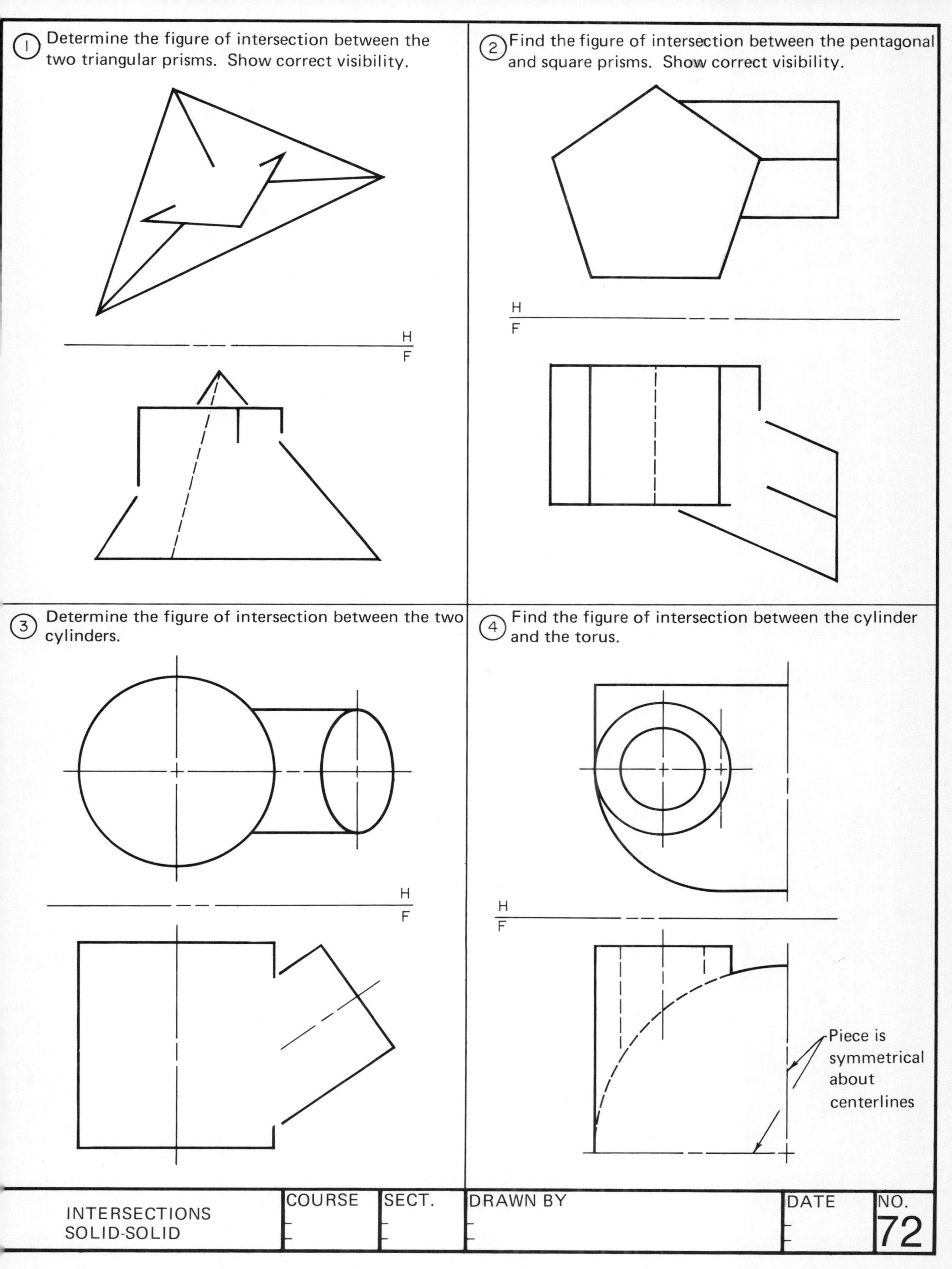

1 Determine the figure of intersection between the two triangular prisms. Show correct visibility.
H
F
2 Find the figure of intersection between the pentagonal and square prisms. Show correct visibility.
H
F
3 Determine the figure of intersection between the two cylinders.
H
F
4 Find the figure of intersection between the cylinder and the torus.
H
F
Piece is symmetrical about centerlines
INTERSECTIONS SOLID-SOLID
COURSE
SECT.
DRAWN BY
DATE
NO.
72

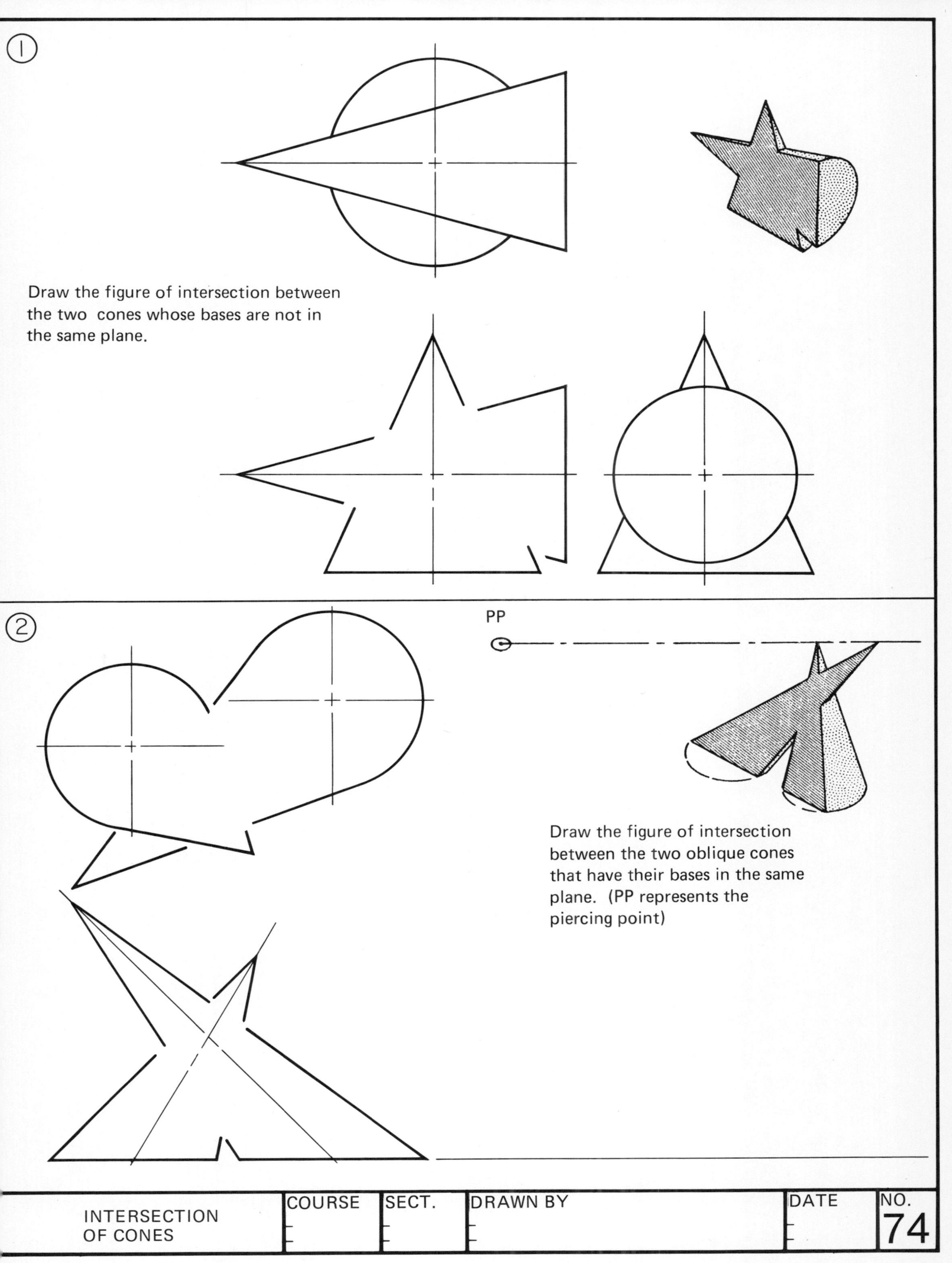

①
Draw the figure of intersection between the two cones whose bases are not in the same plane.
②
PP
Draw the figure of intersection between the two oblique cones that have their bases in the same plane. (PP represents the piercing point)
INTERSECTION OF CONES
COURSE
SECT.
DRAWN BY
DATE
NO.
74

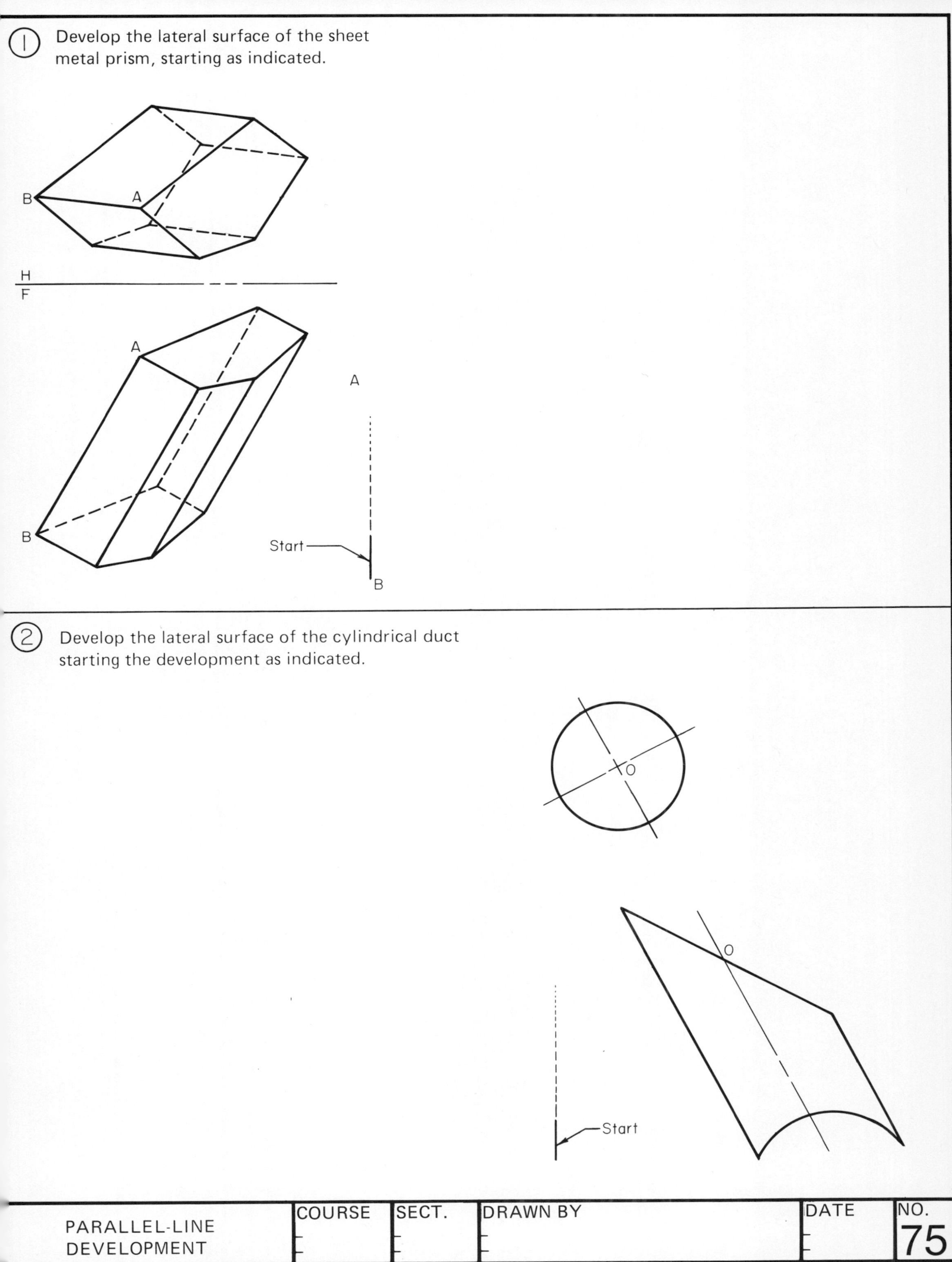
① Develop the lateral surface of the sheet metal prism, starting as indicated.
B
A
H
F
A
B
A
Start
B
② Develop the lateral surface of the cylindrical duct starting the development as indicated.
O
O
Start

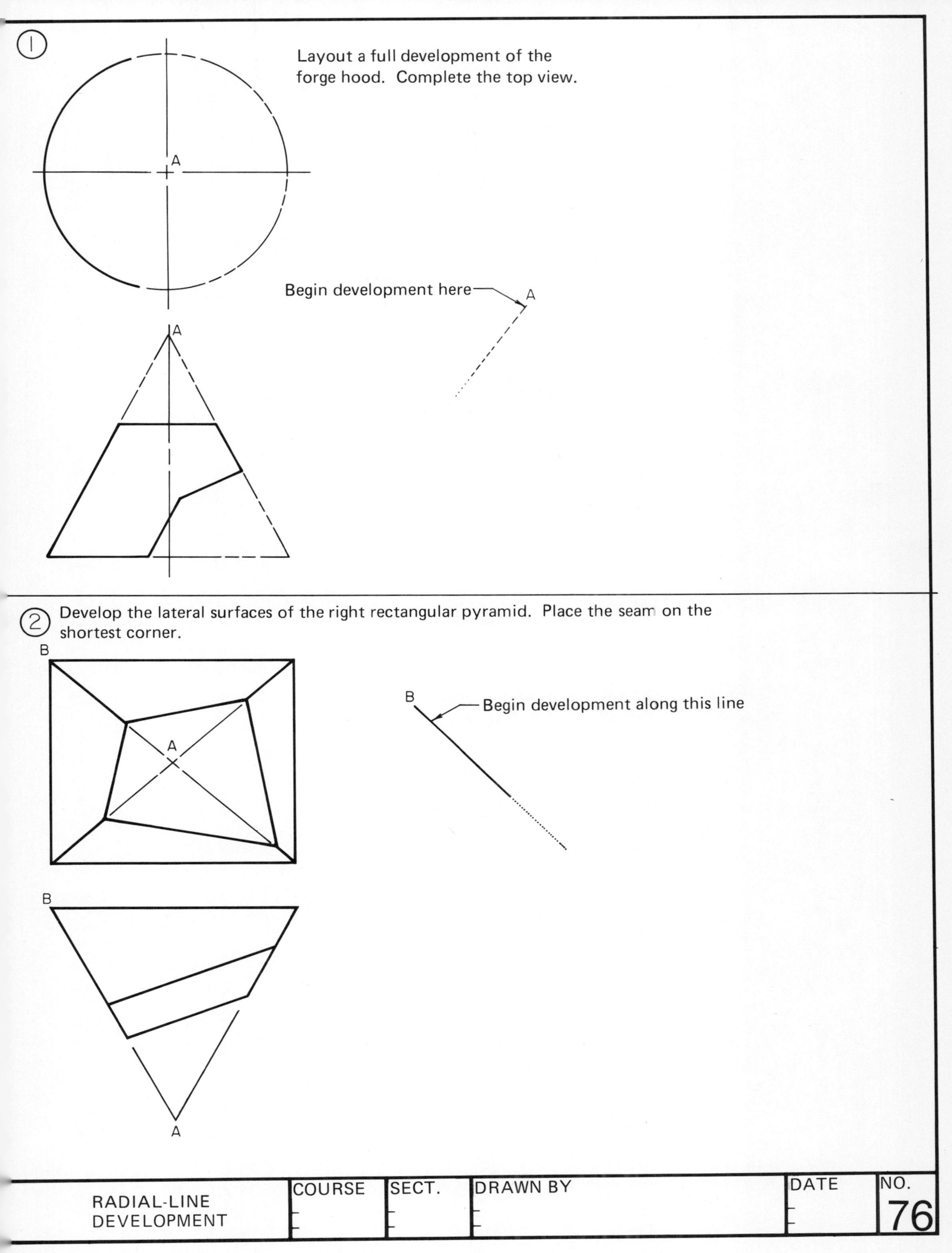
1
Layout a full development of the
forge hood. Complete the top view.
A
Begin development here
A
A
2
Develop the lateral surfaces of the right rectangular pyramid. Place the seam on the
shortest corner.
B
A
B
Begin development along this line
B
A
RADIAL-LINE
DEVELOPMENT
COURSE
SECT.
DRAWN BY
DATE
NO.
76

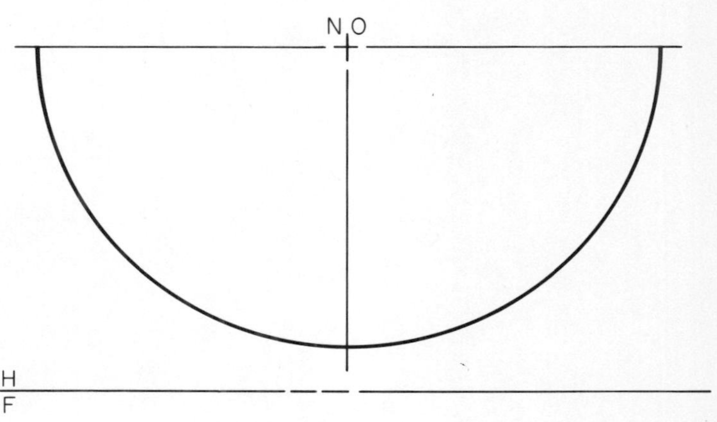

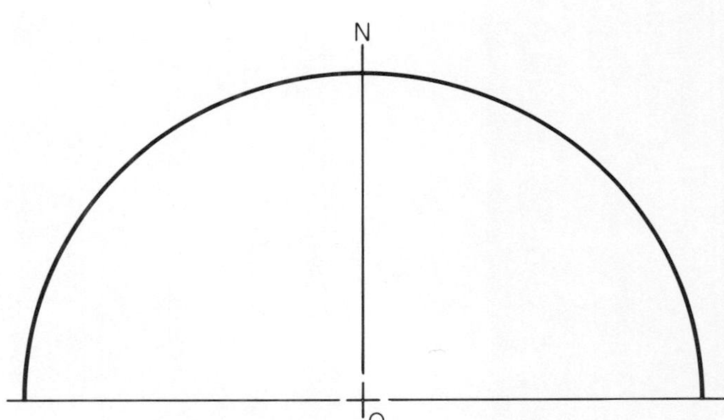

Calculate in Nautical miles, the distance on the earth's surface from point A latitude 65° N, longitude 15° W to point B latitude 45° N, longitude 115° E. Determine the initial and final azimuth bearings of this course.

SPHERICAL NAVIGATION	COURSE	SECT.	DRAWN BY	DATE	NO. 81

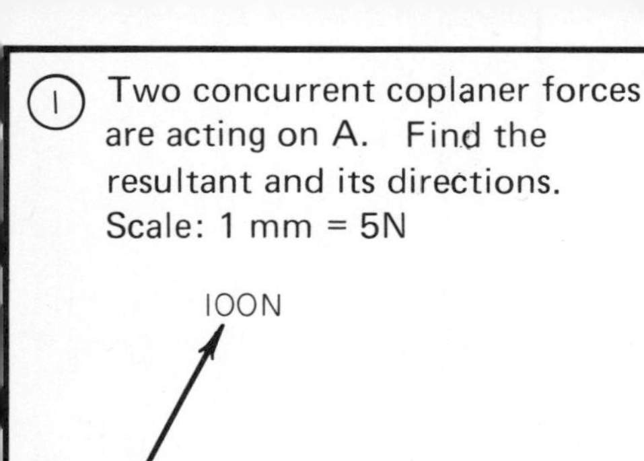

(1) Two concurrent coplaner forces are acting on A. Find the resultant and its directions.
Scale: 1 mm = 5N

(2) Find the resultant of the three concurrent coplaner forces.
Scale: 1 mm = 5N

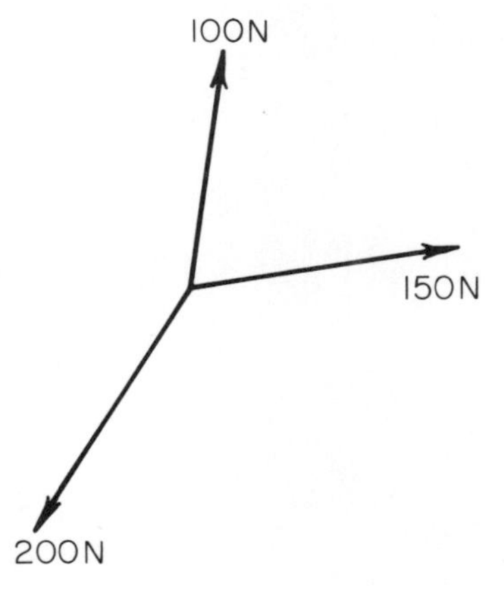

(3) What are the forces acting in the two members supporting the 300N load?
Scale: 1 mm = 5N

C
A
B
300N

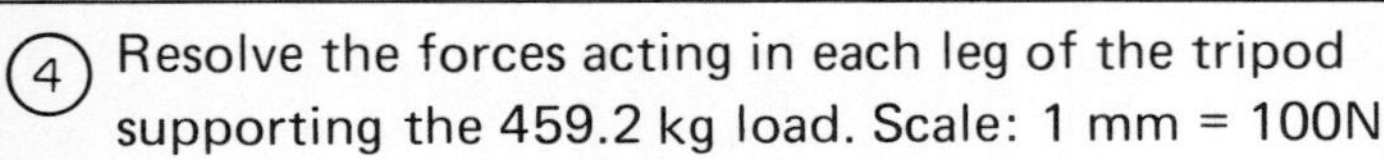

(4) Resolve the forces acting in each leg of the tripod supporting the 459.2 kg load. Scale: 1 mm = 100N

2
0
3
1
H
F

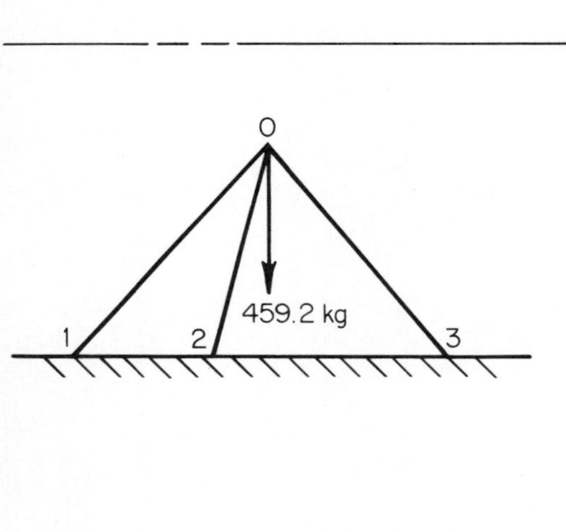

(5) Resolve the three concurrent noncoplaner forces acting on point A. Scale: 1 mm = 5N

A1 = ________ A2 = ________ A3 = ________

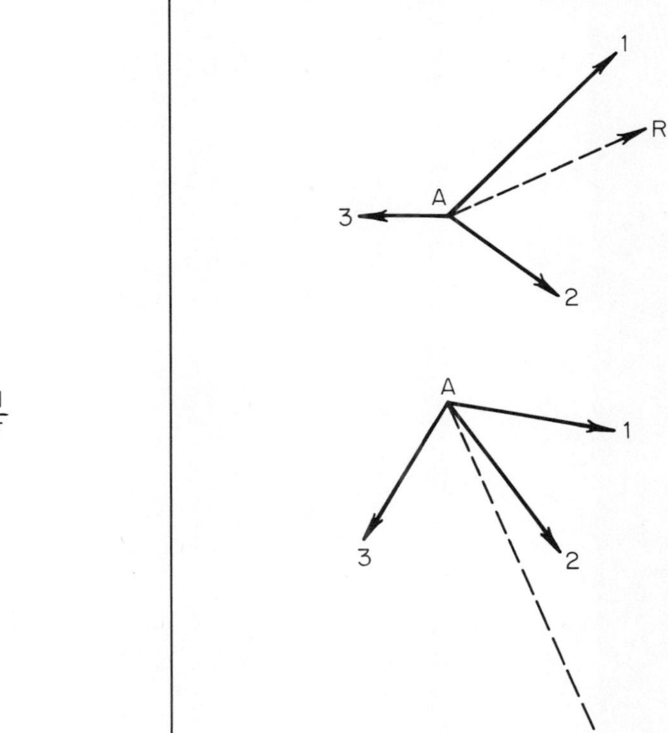

CONCURRENT VECTOR ANALYSIS	COURSE	SECT.	DRAWN BY	DATE	NO. 82

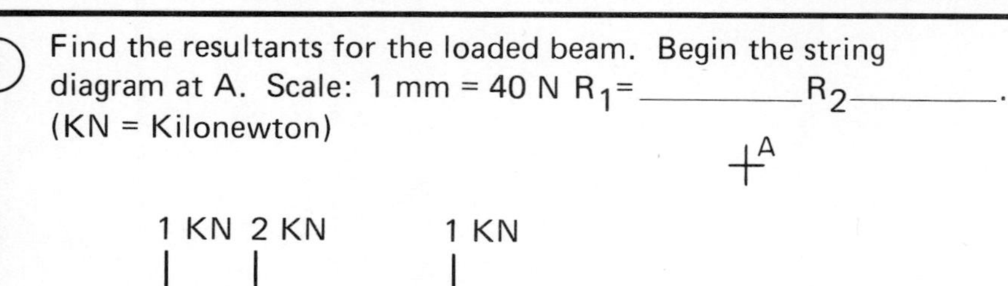

① Find the resultants for the loaded beam. Begin the string diagram at A. Scale: 1 mm = 40 N R_1=__________ R_2__________.
(KN = Kilonewton)

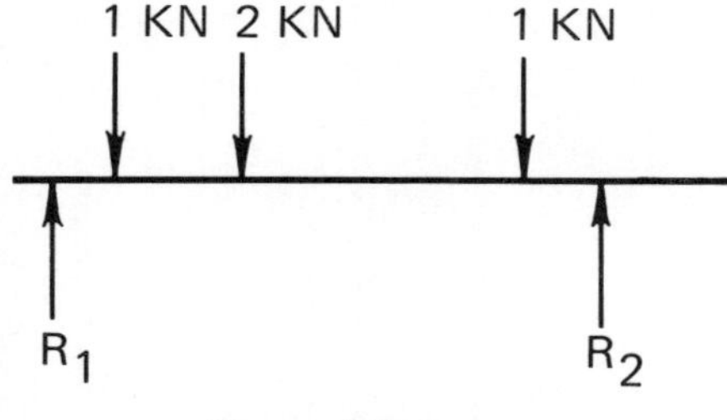

Space Diagram

② Find the resultants for the loaded beam. Begin string diagram at A. Scale: 1 mm = 80 N R_1=__________ R_2=__________.
(KN = Kilonewton)

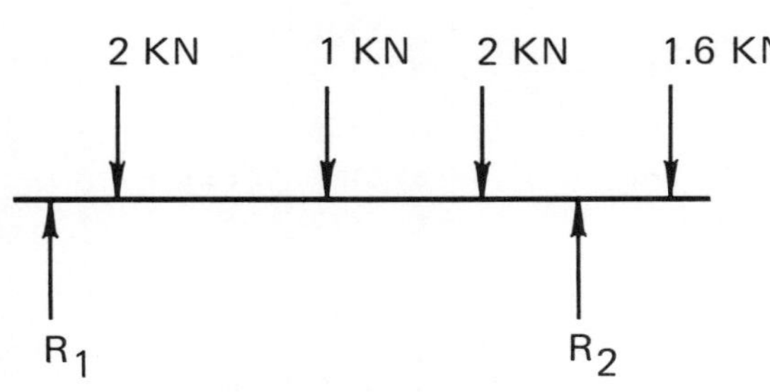

Space Diagram

NONCONCURRENT COPLANER FORCES	COURSE	SECT.	DRAWN BY	DATE	NO. 83

Determine the stresses acting in each member of the Fink truss. Indicate whether each member is in tension (—) or compression (+). Scale: 1 mm = 40 N (KN = Kilonewton) Start stress diagram at A.

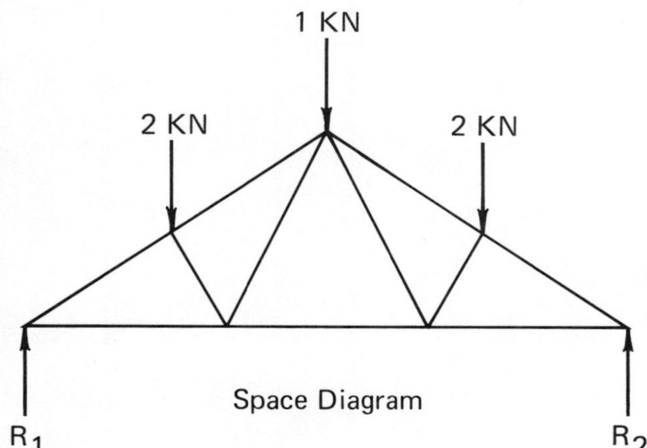

Space Diagram

A

Stress Diagram

Force analysis/Sketches

NONCONCURRENT COPLANER FORCES	COURSE	SECT.	DRAWN BY	DATE	NO. 84

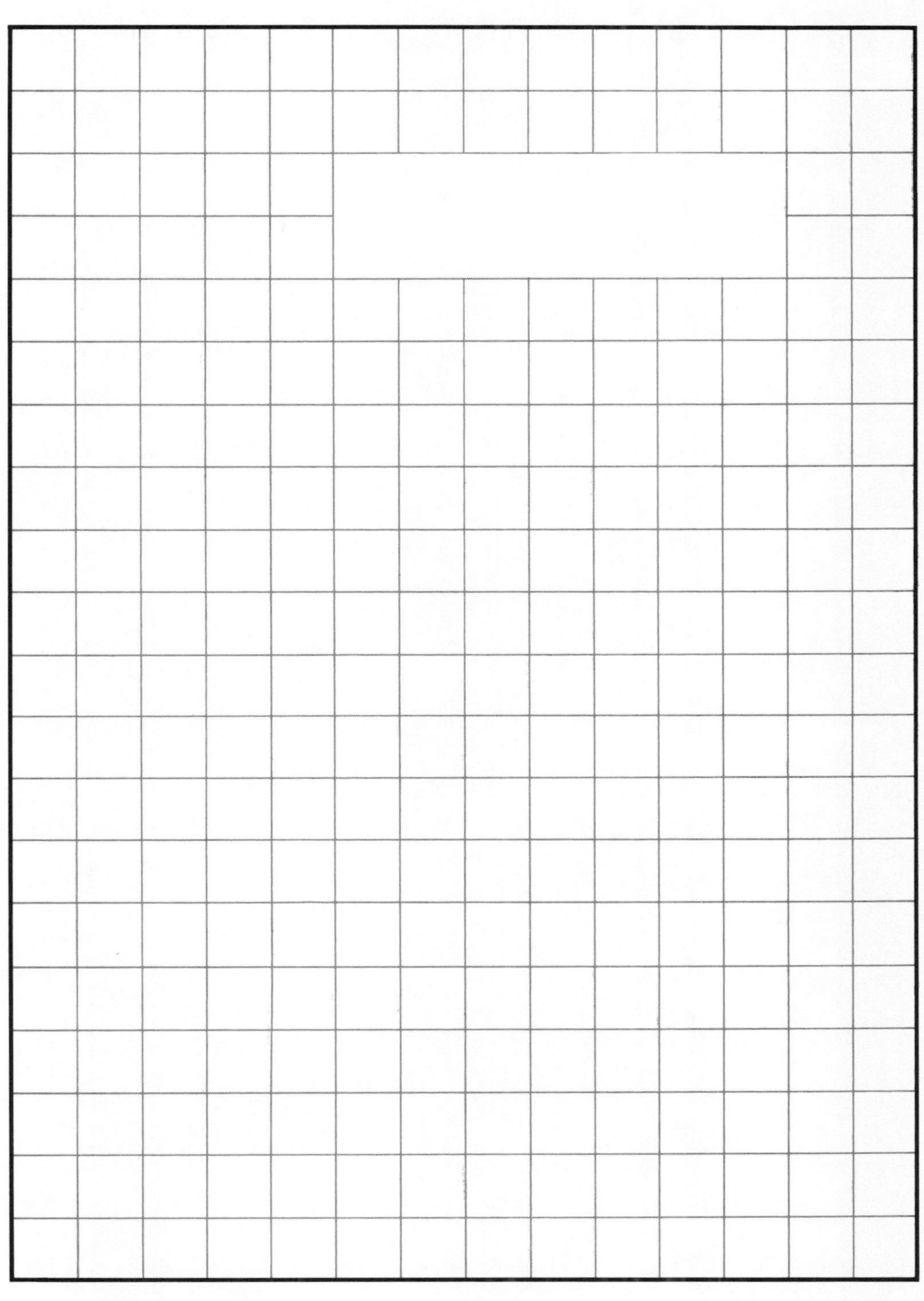

RECTANGULAR COORDINATE GRAPH	COURSE	SECT.	DRAWN BY	DATE	NO. 85

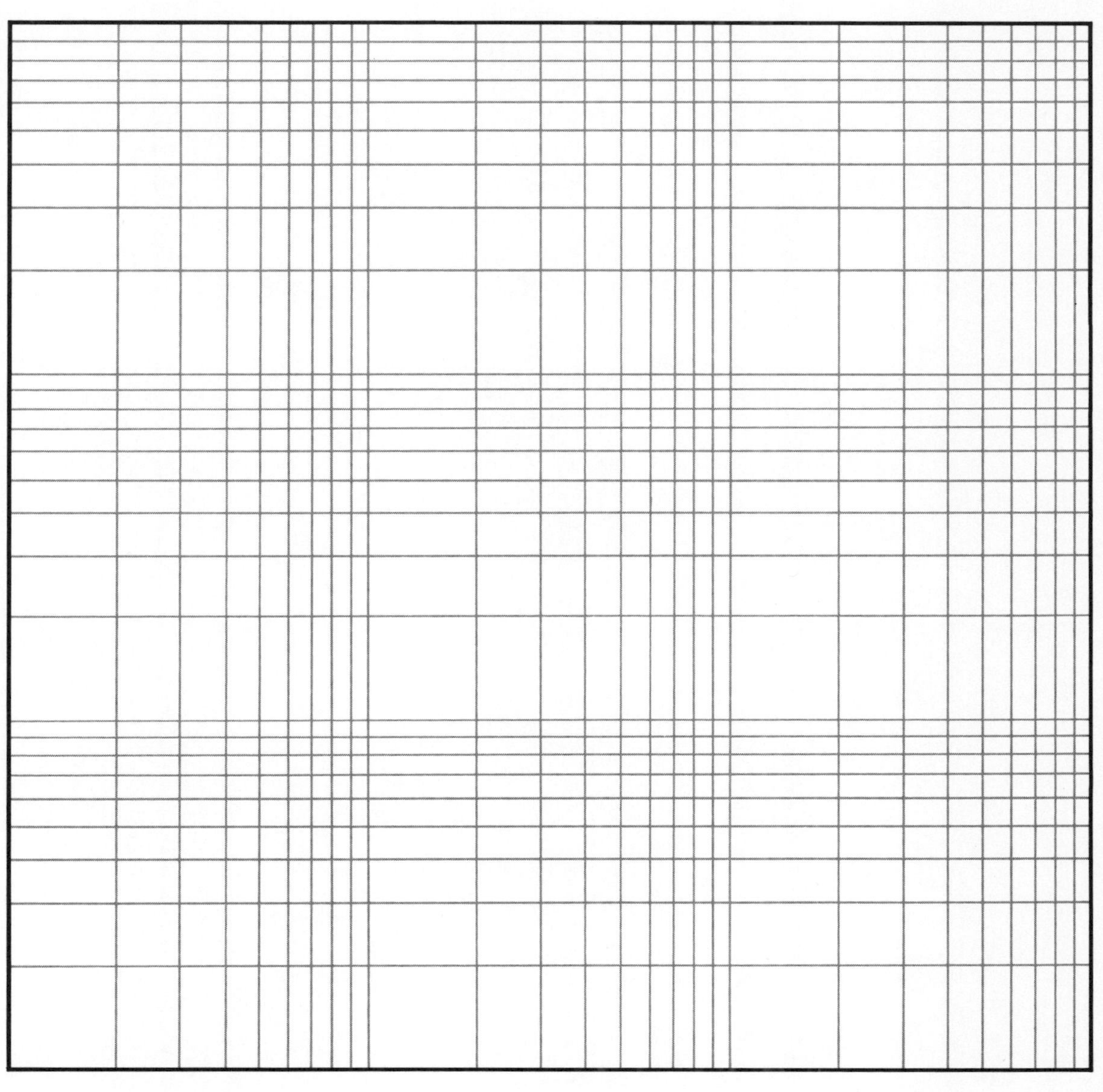

LOGARITHMIC COORDINATE GRAPH	COURSE	SECT.	DRAWN BY	DATE	NO. 87

INTRODUCTION TO COMPUTER GRAPHICS

Computer graphics fundamentals are based upon defining points and lines within a mathematical coordinate system. The problems in the following section are designed to present fundamental concepts and provide exercises in defining objects for computer graphics applications. To assist the student in completing the problem material, a basic introduction regarding two and three dimensional coordinate systems has been provided. Information regarding generic command language to help the student develop plotting subroutines has also been included below.

Two-dimensional coordinate systems-use a horizontal 'X' axis (abscissa) and a vertical 'Y' axis (ordinate) to define points on the X,Y plane. Scale values for a two-dimensional coordinate system follow the same rules that apply to mathematical coordinate graphs. After points have been defined as X,Y values, a sequential series of written commands (called subroutines) direct the computer to plot lines based upon the X,Y values. The written commands required to create subroutines for two-dimensional problem material in this section take the following generic form (although other forms may be used with your instructor's permission).

COMMAND FORM	DEFINITION
CALL MOVE (X,Y)	Pen moves from its present position to new X,Y coordinates indicated (No line is drawn).
CALL DRAW (X,Y,2)	A line is drawn from present pen position to new X,Y coordinates indicated. Numeral 2 indicates a visible line is drawn.
CALL HOLE (X,Y,DIA.,2)	Creates hole with X,Y coordinates as center point. DIA. is the diameter of hole. Numeral 2 indicates a visible line is drawn.
CALL ARC (X,Y,RAD.,BA,EA,2)	Draws an arc through a 90° angle, with X,Y coordinates as the centerpoint. RAD. is the radius of arc. BA is the beginning angle of arc and EA is the ending angle of arc. BA and EA values may be 0,90,180, or 270 (0 is at top of sheet. Values increase clockwise). Arcs may be plotted clockwise or counter-clock-wise depending on BA and EA values. After arc is drawn, pen position is at end of arc. Numeral 2 indicates a visible line is drawn.

Three-dimensional coordinate systems-use an X axis and Y axis as in two-dimensional coordinate systems, but also include a third 'Z' axis placed perpendicular to the X,Y plane. Points may now take a position away from or off the X,Y plane and are defined in space having an X, Y and Z value. Z values are positive (+) in front of the X, Y plane and become negative (–) in a position behind the X, Y plane. The written commands required to create subroutines for three-dimensional problem material in this section take the following generic form(although others may be used with your instructor's permission).

COMMAND FORM	DEFINITION
CALL MOVE (X,Y,Z)	Pen moves from its present position to the new X,Y,Z coordinates indicated (No line is drawn).
CALL DRAW (X,Y,Z,2)	A line is drawn from present pen position to new X,Y,Z coordinates indicated. Numeral 2 indicates a visible line is drawn.

① Digitize the single-view drawing of the COVER PLATE by defining the X and Y coordinate values of the indicated points. Define the coordinate values as decimal units. Point A is the origin of the X and Y axes and has a scale value of 0.0,0.0

② Define the X and Y coordinate values of the indicated points on the layout of the STEP SUPPORT (use decimal units). Then, create a subroutine using the necessary commands for two-dimensional drawings listed in the introduction of this section. Point A is the origin of the X and Y axes and has a scale value of 0.0,0.0. Indicate any negative (–) X,Y values in your subroutine. Write the subroutine to plot the drawing in a clockwise direction.

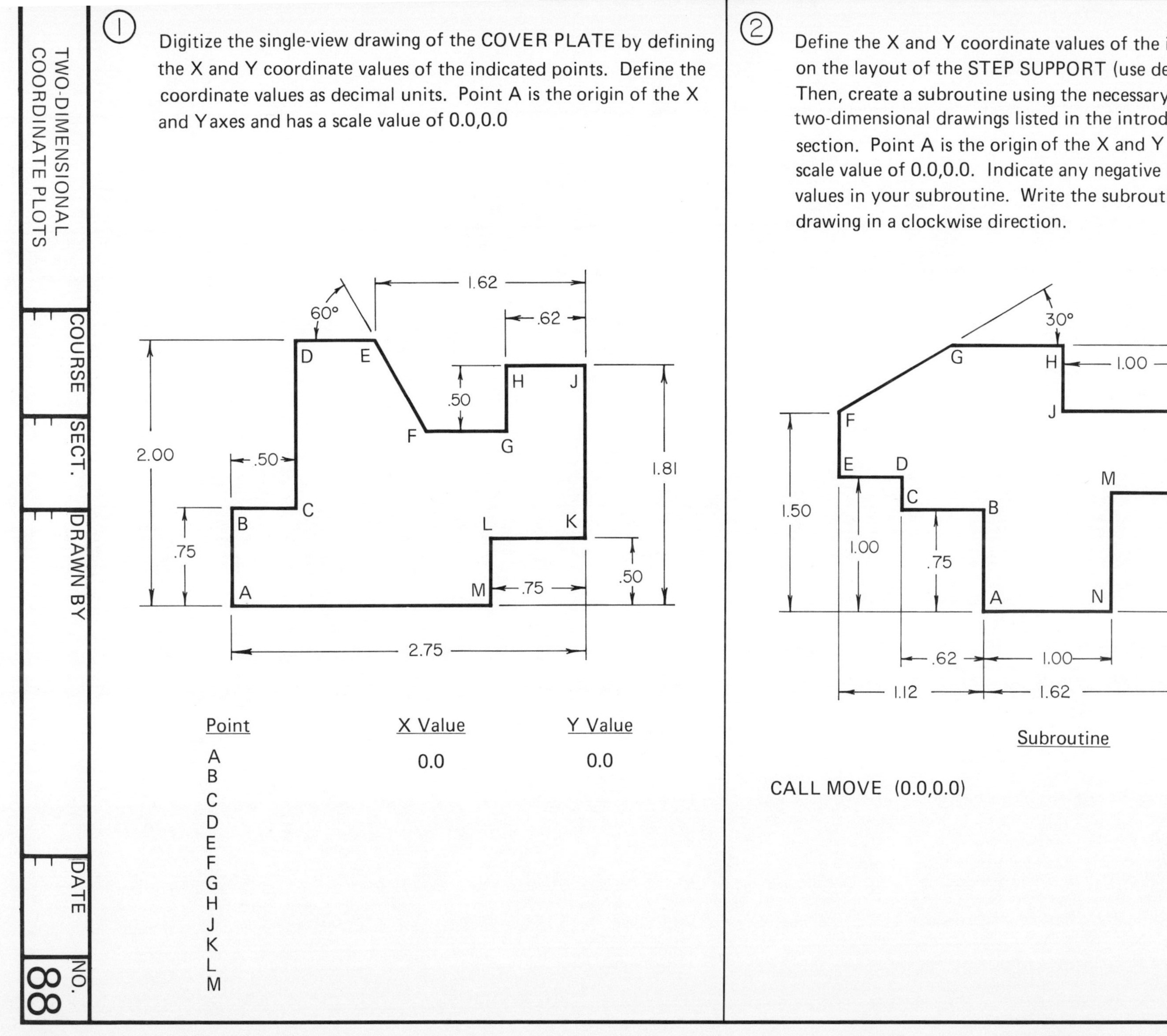

Point	X Value	Y Value
A	0.0	0.0
B		
C		
D		
E		
F		
G		
H		
J		
K		
L		
M		

Subroutine

CALL MOVE (0.0,0.0)

TWO-DIMENSIONAL COORDINATE PLOTS	COURSE	SECT.	DRAWN BY	DATE	NO. 88

Create a subroutine using the necessary commands for two-dimensional drawings to plot a full-scale layout of the SHAFT GUIDE. Define any necessary points as full scale decimal-inch coordinates. Indicate any negative (–) X, Y coordinates in your subroutine. Point A has been referenced as the origin of the X and Y axes and has a scale value of 0.0,0.0. Write the subroutine to plot the layout in a clockwise direction.

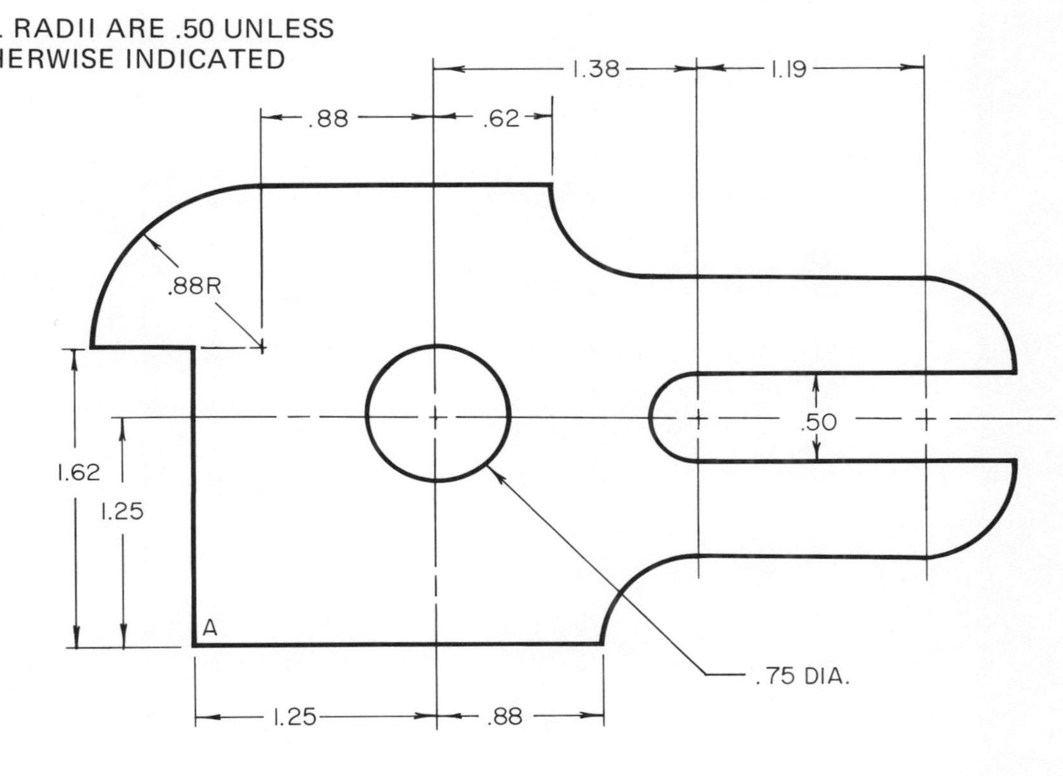

Subroutine

TWO-DIMENSIONAL COORDINATE PLOT	COURSE	SECT.	DRAWN BY	DATE	NO. 89

Digitize the full size wire frame diagram of the BEVEL BASE, by defining the X,Y and Z values for each lettered point. (Use decimal units). Then, create a subroutine using the necessary commands for three-dimensional drawings listed in the introduction of this section. Point A is the origin of the X,Y and Z axes and has a scale value of 0.0, 0.0, 0.0. Do not retrace any lines that have already been plotted in your subroutine (Refer to the two view orthographic drawing for dimensional information).

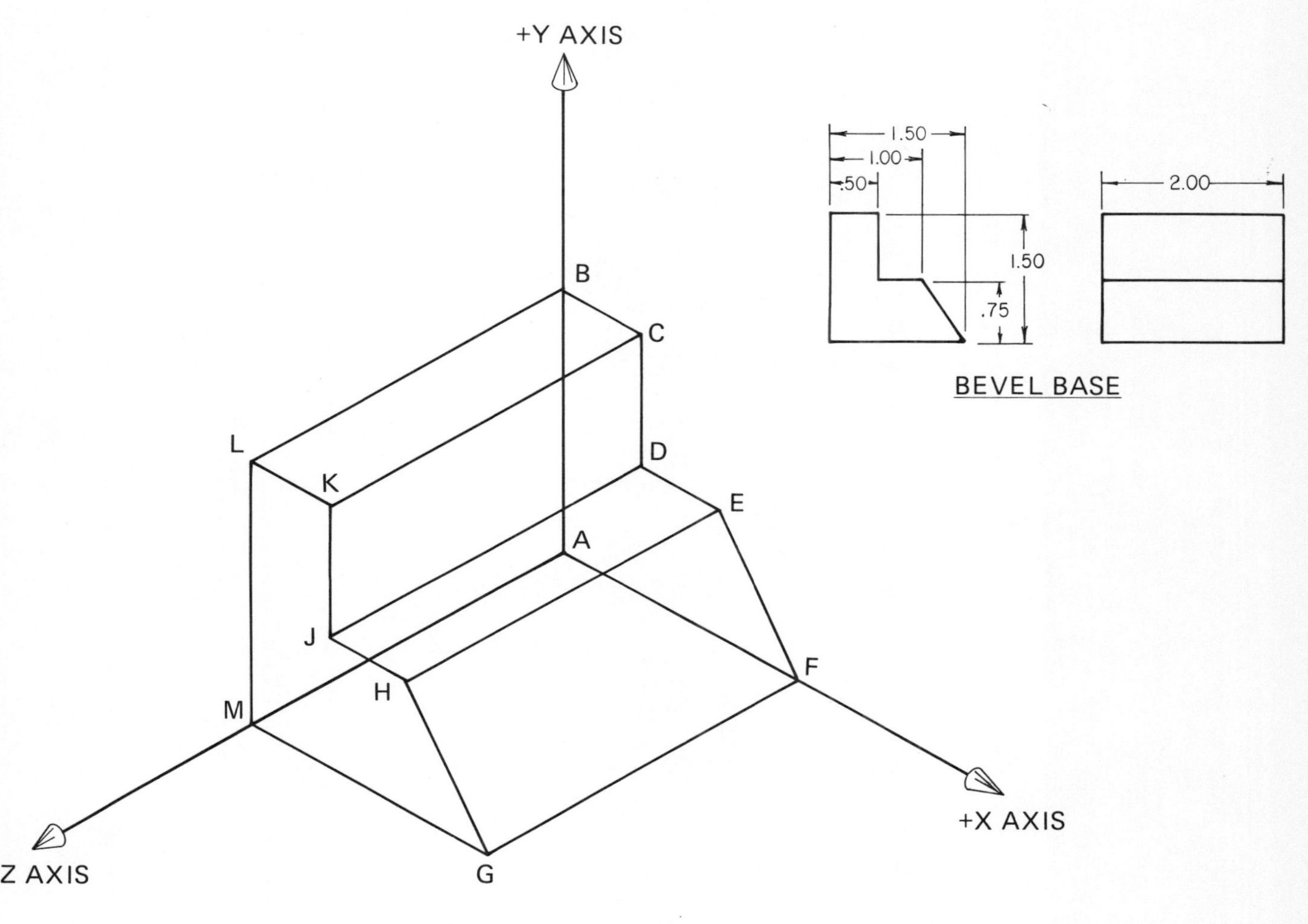

BEVEL BASE

Point	X,Y,Z Value	Subroutine
A		
B		
C		
D		
E		
F		
G		
H		
J		
K		
L		
M		

THREE-DIMENSIONAL COORDINATE PLOT	COURSE	SECT.	DRAWN BY	DATE	NO. 90

Using the listed subroutine, plot the indicated wire frame object on the coordinate drawing below. Coordinate values are given in the subroutine as actual decimal-inch units. The X, Y and Z axes are drawn full size.

Subroutine

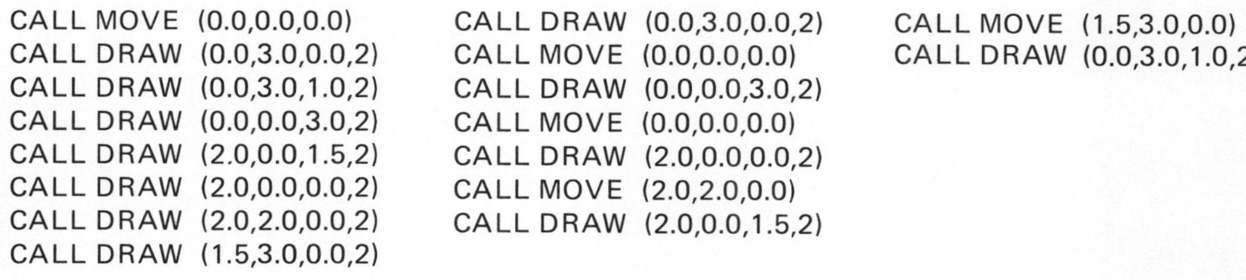

```
CALL MOVE  (0.0,0.0,0.0)
CALL DRAW  (0.0,3.0,0.0,2)
CALL DRAW  (0.0,3.0,1.0,2)
CALL DRAW  (0.0,0.0,3.0,2)
CALL DRAW  (2.0,0.0,1.5,2)
CALL DRAW  (2.0,0.0,0.0,2)
CALL DRAW  (2.0,2.0,0.0,2)
CALL DRAW  (1.5,3.0,0.0,2)
CALL DRAW  (0.0,3.0,0.0,2)
CALL MOVE  (0.0,0.0,0.0)
CALL DRAW  (0.0,0.0,3.0,2)
CALL MOVE  (0.0,0.0,0.0)
CALL DRAW  (2.0,0.0,0.0,2)
CALL MOVE  (2.0,2.0,0.0)
CALL DRAW  (2.0,0.0,1.5,2)
CALL MOVE  (1.5,3.0,0.0)
CALL DRAW  (0.0,3.0,1.0,2)
```

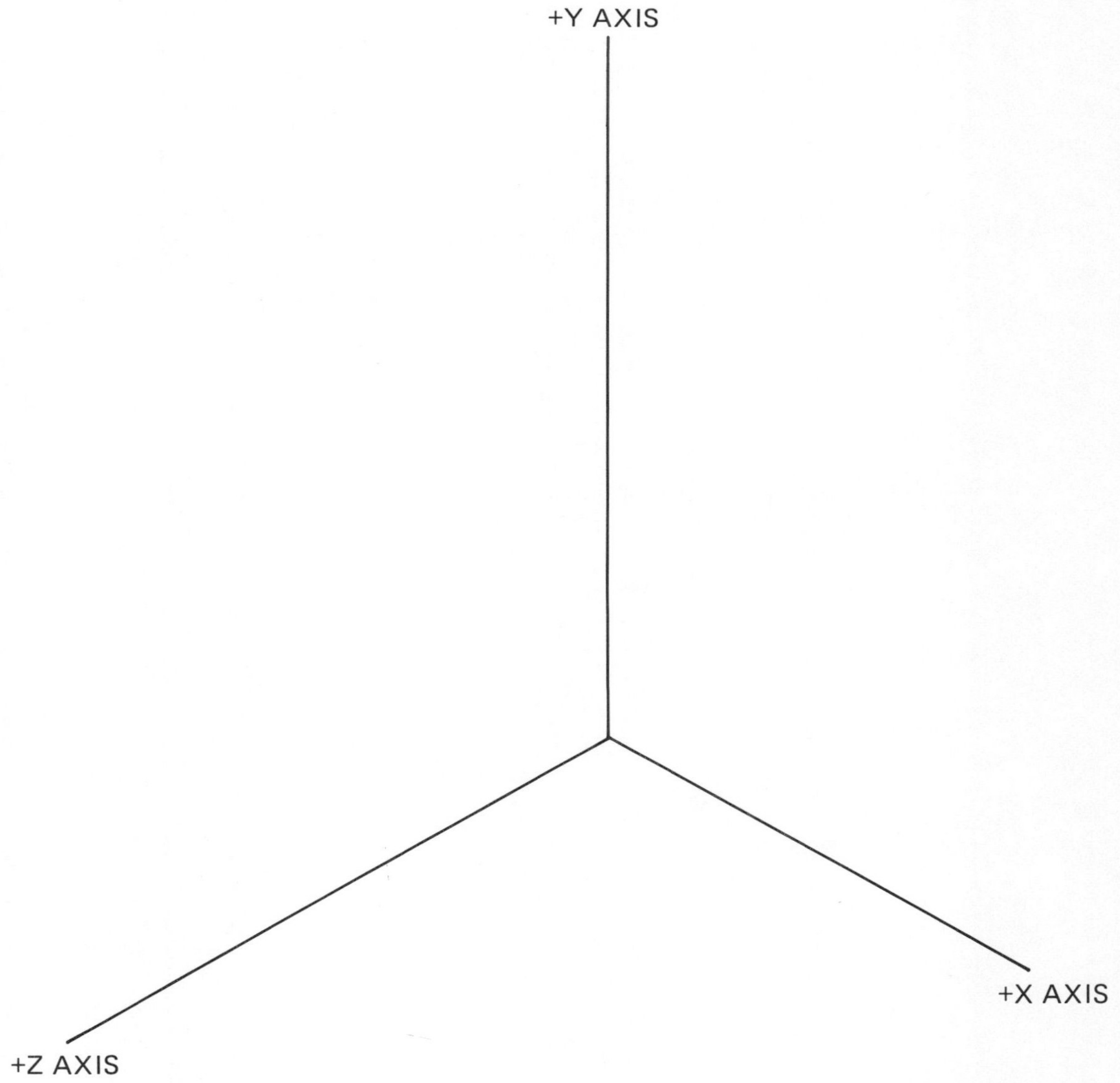

+Y AXIS

+X AXIS

+Z AXIS

THREE-DIMENSIONAL SUBROUTINE PLOT	COURSE	SECT.	DRAWN BY	DATE	NO. 91

①

②

③

④

⑤

⑥

VERTICAL LETTERING	COURSE	SECT.	DRAWN BY	DATE	NO. 3

1
2
3
4
5
6
VERTICAL LETTERING
COURSE
SECT.
DRAWN BY
DATE
NO.
3

1

2

3

4

5

6

INCLINED LETTERING	COURSE	SECT.	DRAWN BY	DATE	NO. 7

(1)

(2)

(3)

(4)

(5)

(6)

INCLINED LETTERING	COURSE	SECT.	DRAWN BY	DATE	NO. 7

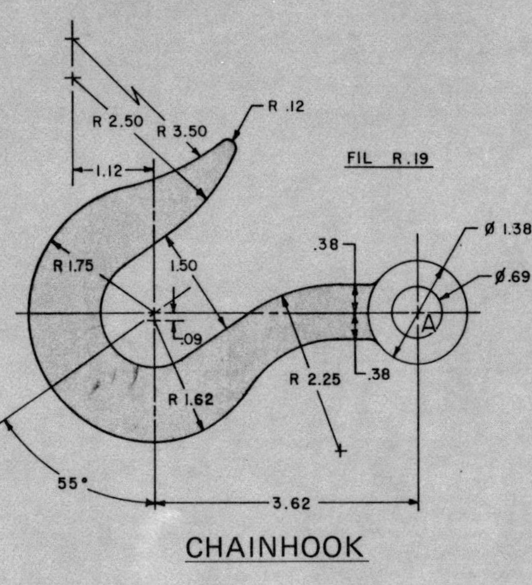

CHAINHOOK

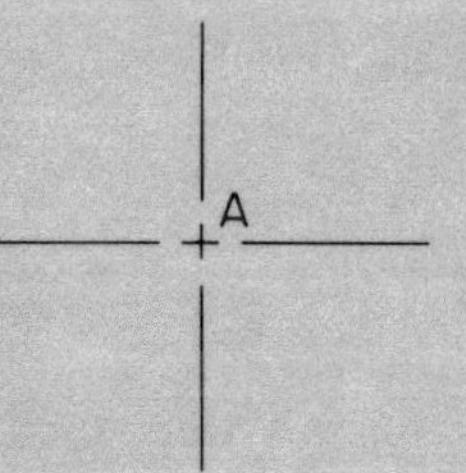

Make a full size drawing
of the Chainhook.

GEOMETRIC
CONSTRUCTIONS

DRAWN BY

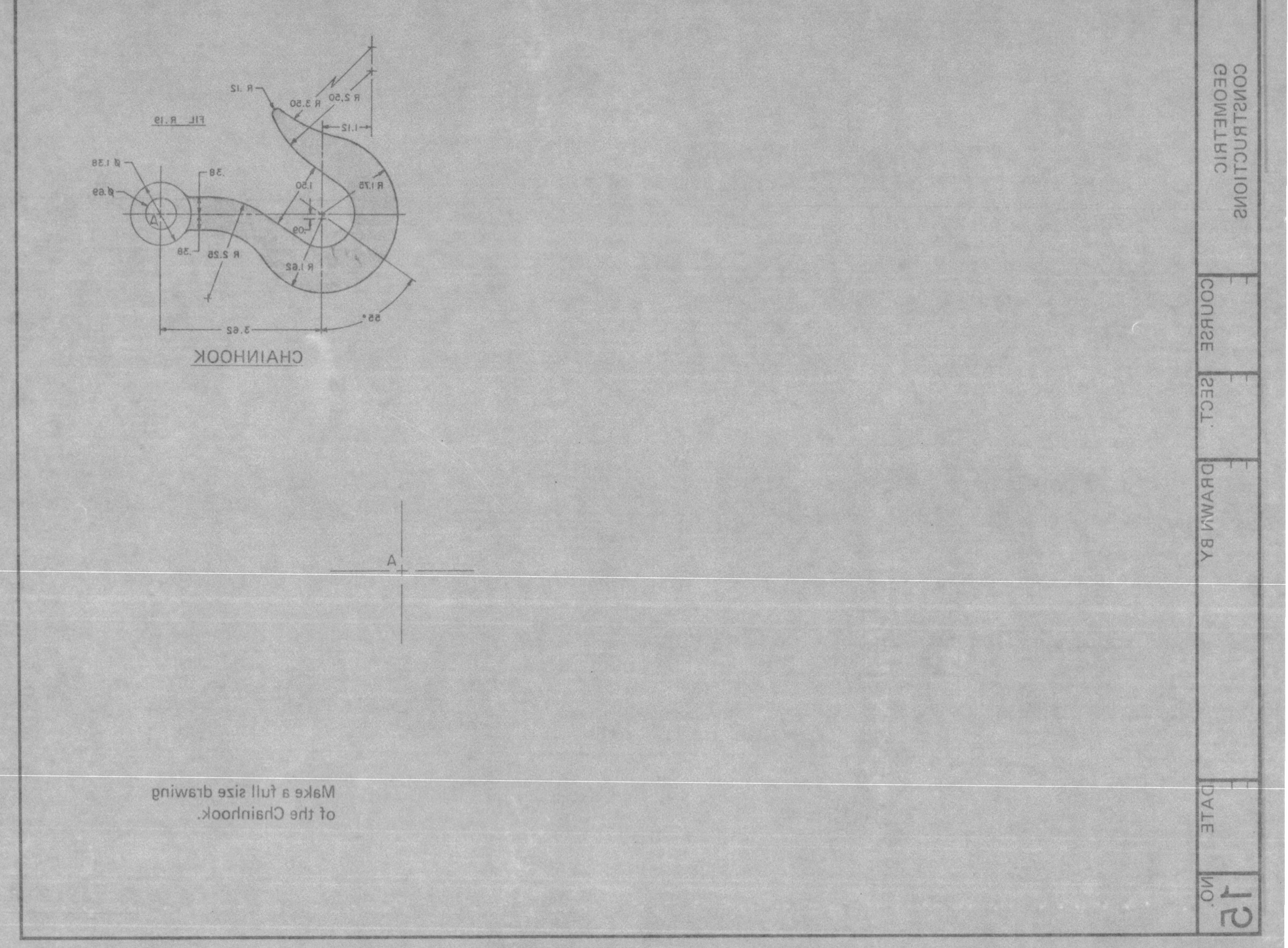

CHAINHOOK
Make a full size drawing
of the Chainhook.
A
GEOMETRIC
CONSTRUCTIONS
COURSE
SECT.
DRAWN BY
DATE
NO.
15

Make a single view drawing of the AIR SHUTTER. Design the slot for a M6.3 x 1 machine screw to permit travel equal to the air opening. (Allow a clearance of .7 mm for the fastener to pass through the slot.) The diameter of the center hole is 16.5 mm. The AIR SHUTTER is manufactured from 3.0 mm steel sheet stock.

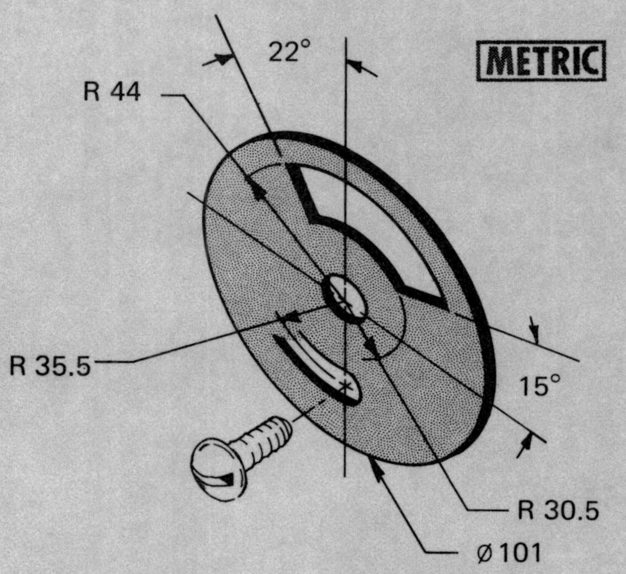

AIR SHUTTER

GEOMETRIC CONSTRUCTIONS	COURSE	SECT.	DRAWN BY	DATE	NO. 16

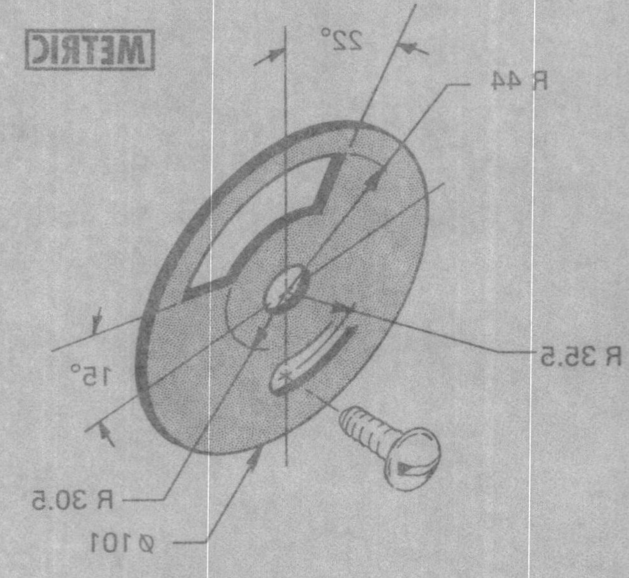

AIR SHUTTER

Make a single view drawing of the AIR SHUTTER. Design the slot for a M6.3 x 1 machine screw to permit travel equal to the air opening. (Allow a clearance of .7 mm for the fastener to pass through the slot.) The diameter of the center hole is 16.5 mm. The AIR SHUTTER is manufactured from 3.0 mm steel sheet stock.

GEOMETRIC CONSTRUCTIONS	COURSE	SECT.	DRAWN BY	DATE	NO. 16

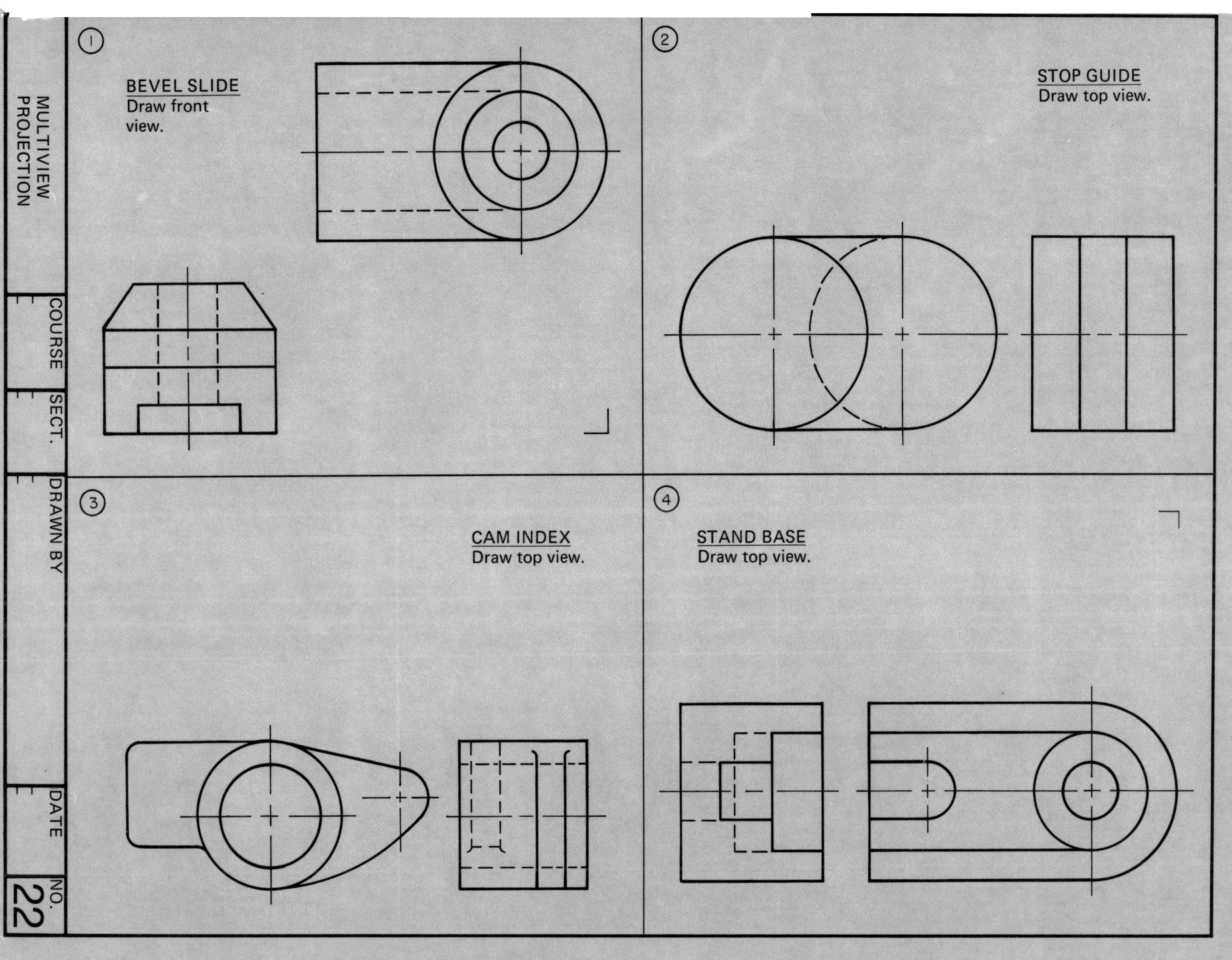
1
BEVEL SLIDE
Draw front view.
2
STOP GUIDE
Draw top view.
3
CAM INDEX
Draw top view.
4
STAND BASE
Draw top view.
MULTIVIEW PROJECTION
COURSE
SECT.
DRAWN BY
DATE
NO.
22

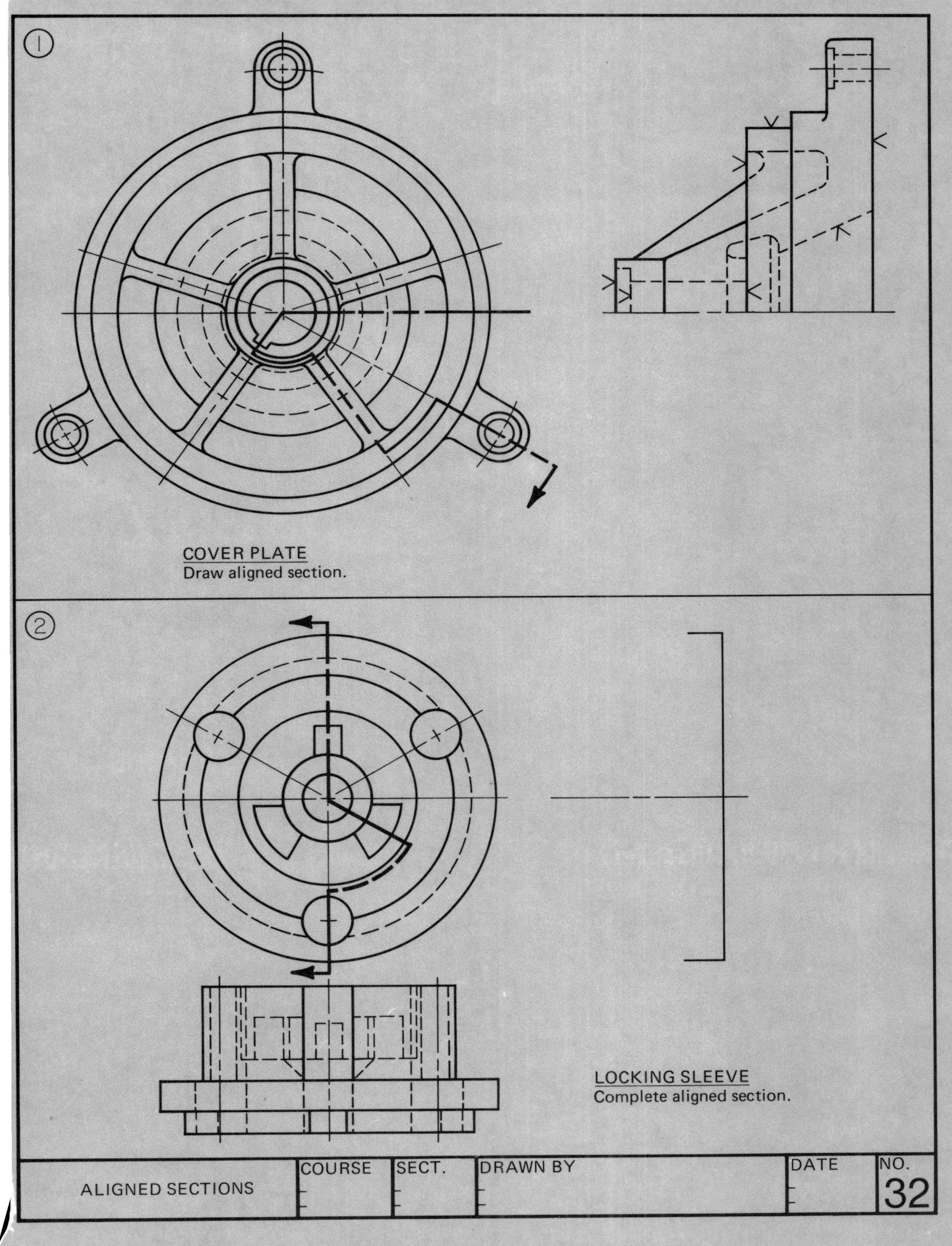
1
COVER PLATE
Draw aligned section.
2
LOCKING SLEEVE
Complete aligned section.
ALIGNED SECTIONS
COURSE
SECT.
DRAWN BY
DATE
NO.
32

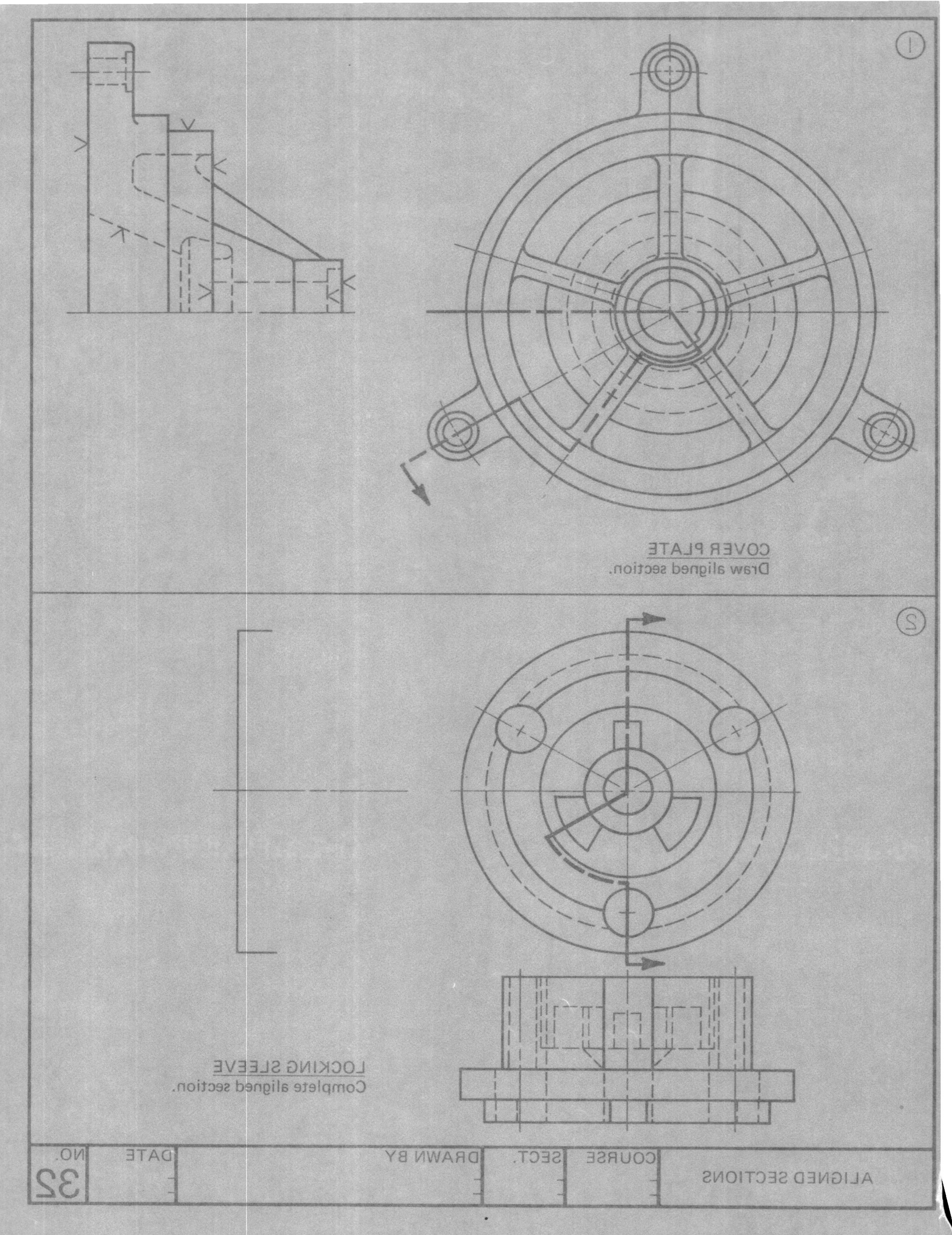

1
COVER PLATE
Draw aligned section.
2
LOCKING SLEEVE
Complete aligned section.
ALIGNED SECTIONS
COURSE
SECT.
DRAWN BY
DATE
NO.
32

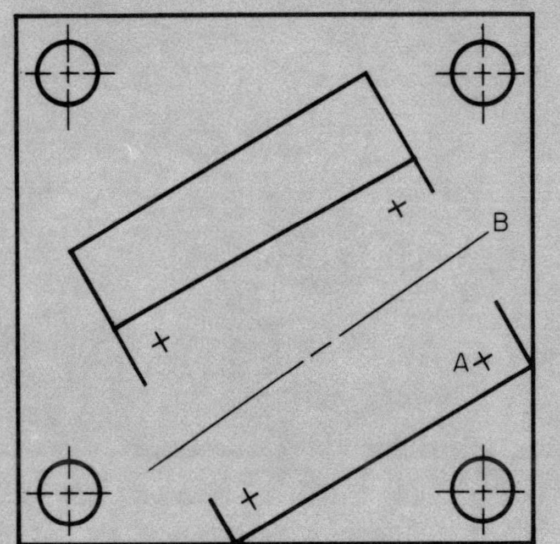

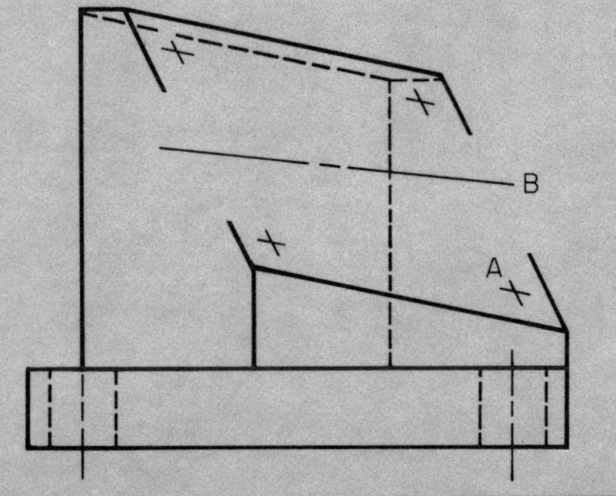

DOVETAIL SUPPORT

Draw sufficient auxiliary views to indicate the true size of the oblique surface A. Locate the true position of the four centerlines (indicated in the front and top views) for dimensioning purposes. Draw an additional auxiliary view to show the true size and shape of the dovetail slide (see detail) which is positioned along centerline B. Complete the top and front views of the dovetail slide.

60°
.19
.75

DOVETAIL SLIDE DETAIL

(Not to scale)

SECONDARY AUXILIARY VIEWS	COURSE	SECT.	DRAWN BY	DATE	NO. 40

DOVETAIL SUPPORT

Draw sufficient auxiliary views to indicate the true size of the oblique surface A. Locate the true position of the four center-lines (indicated in the front and top views) for dimensioning purposes. Draw an additional auxiliary view to show the true size and shape of the dovetail slide (see detail) which is positioned along center-line B. Complete the top and front views of the dovetail slide.

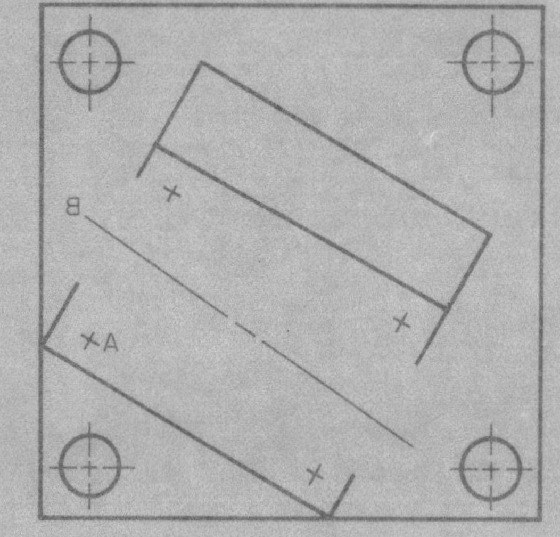

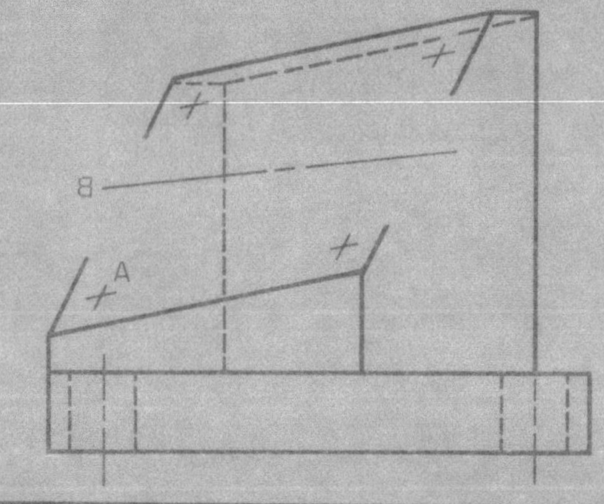

60°
.19
.75

DOVETAIL SLIDE DETAIL
(Not to scale)

SECONDARY AUXILIARY VIEWS	COURSE	SECT.	DRAWN BY	DATE	NO. 40

1

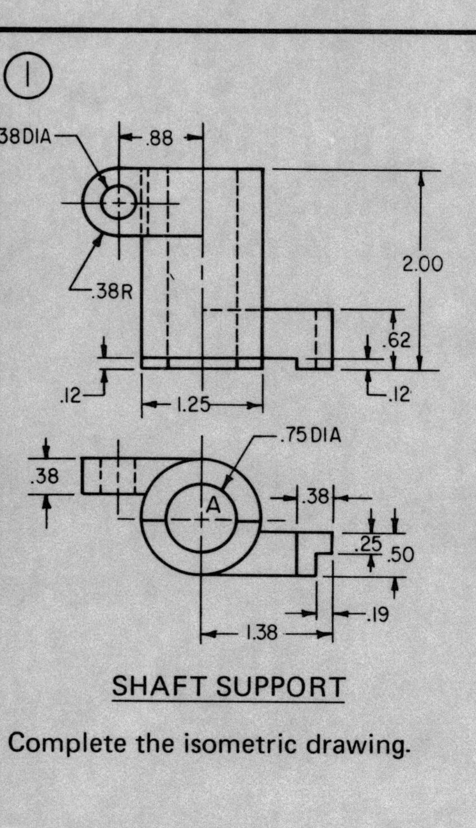

SHAFT SUPPORT

Complete the isometric drawing.

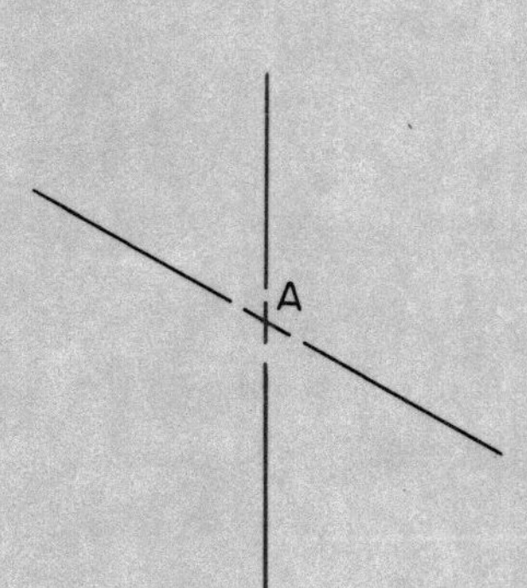

2

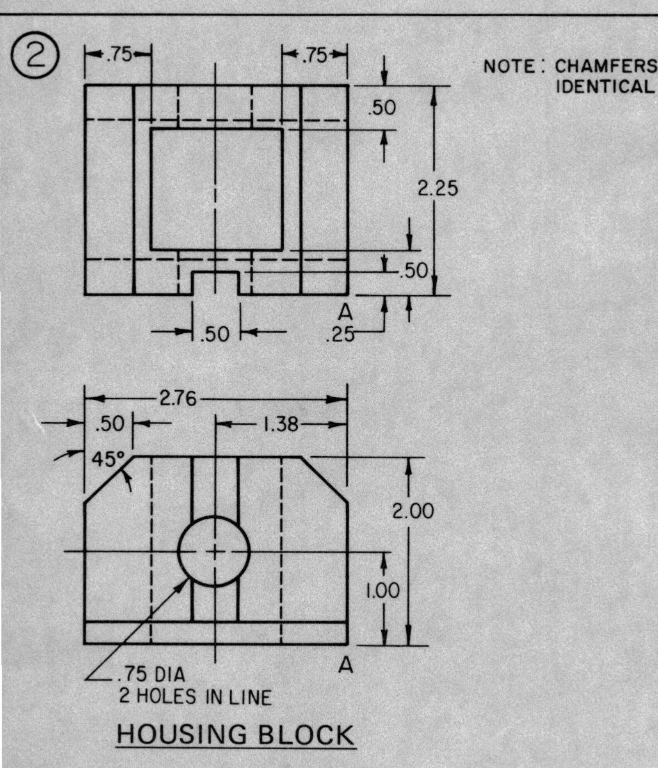

HOUSING BLOCK

NOTE: CHAMFERS ARE IDENTICAL.

Complete the isometric drawing.

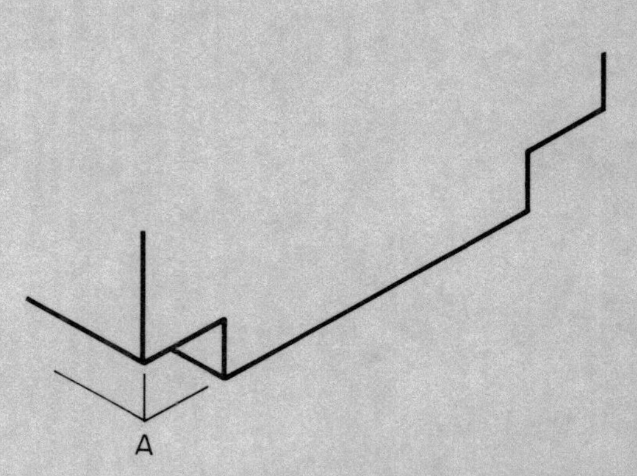

ISOMETRIC DRAWING

DRAWN BY

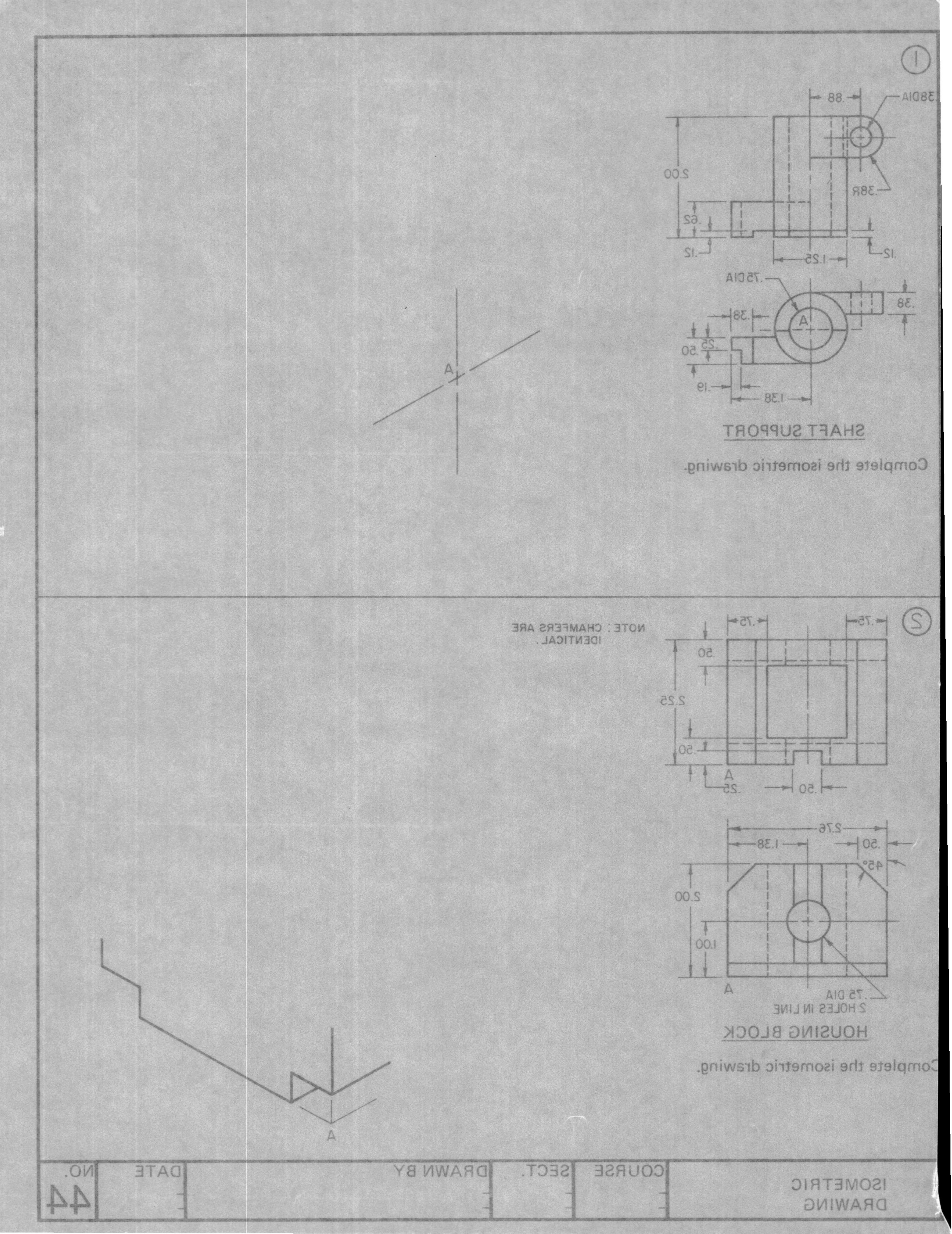
1
.88
.38DIA
2.00
.38R
.62
.12
1.25
.12
.75DIA
.38
.38
A
.25
.50
.19
1.38
SHAFT SUPPORT
Complete the isometric drawing.
A
2
.75
.75
NOTE: CHAMFERS ARE IDENTICAL.
.50
2.25
.50
A
.25
.50
2.76
1.38
.50
45°
2.00
1.00
A
.75 DIA
2 HOLES IN LINE
HOUSING BLOCK
Complete the isometric drawing.
A
ISOMETRIC DRAWING
COURSE
SECT.
DRAWN BY
DATE
NO.
44

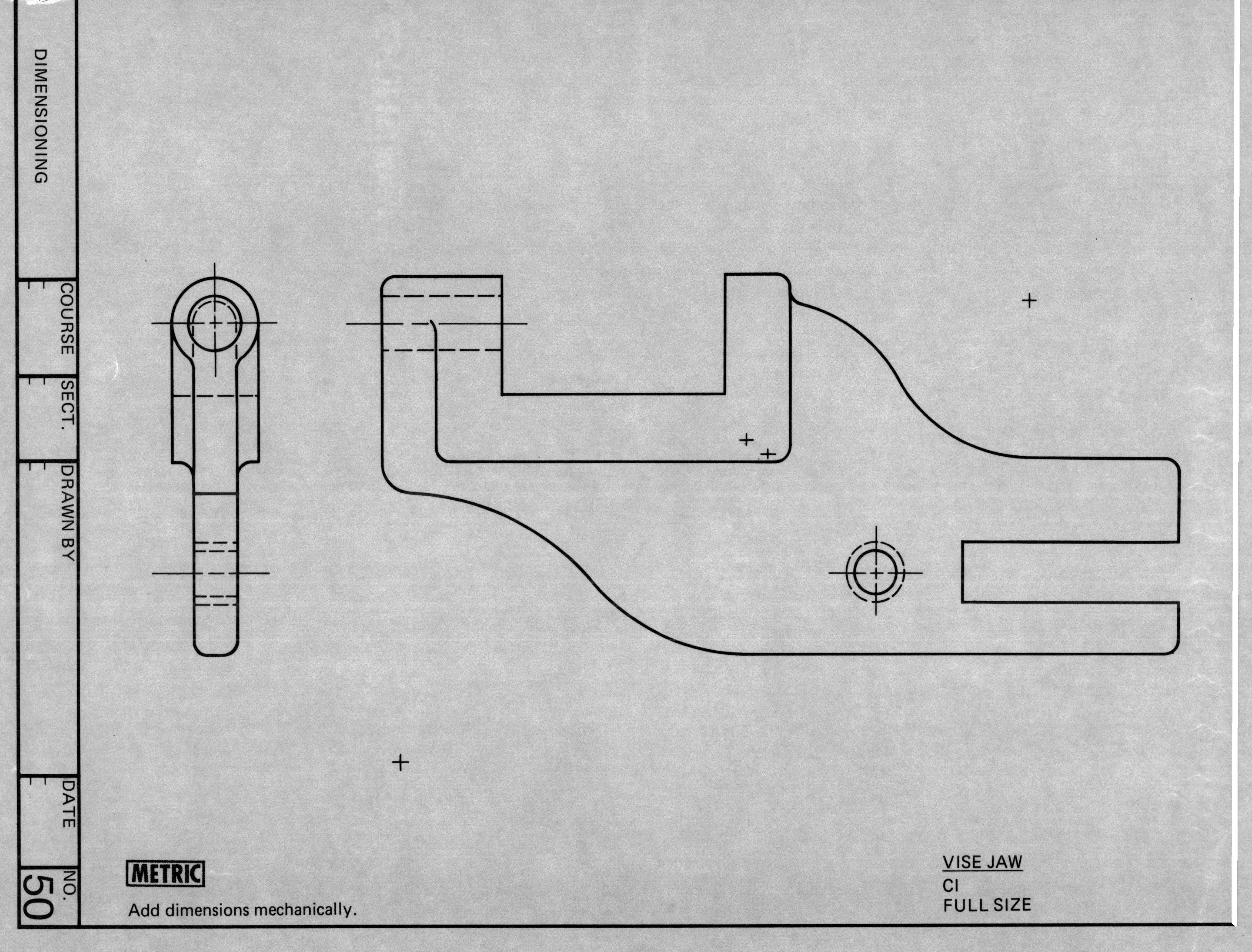

METRIC
Add dimensions mechanically.
VISE JAW
CI
FULL SIZE
DIMENSIONING
COURSE
SECT.
DRAWN BY
DATE
NO.
50

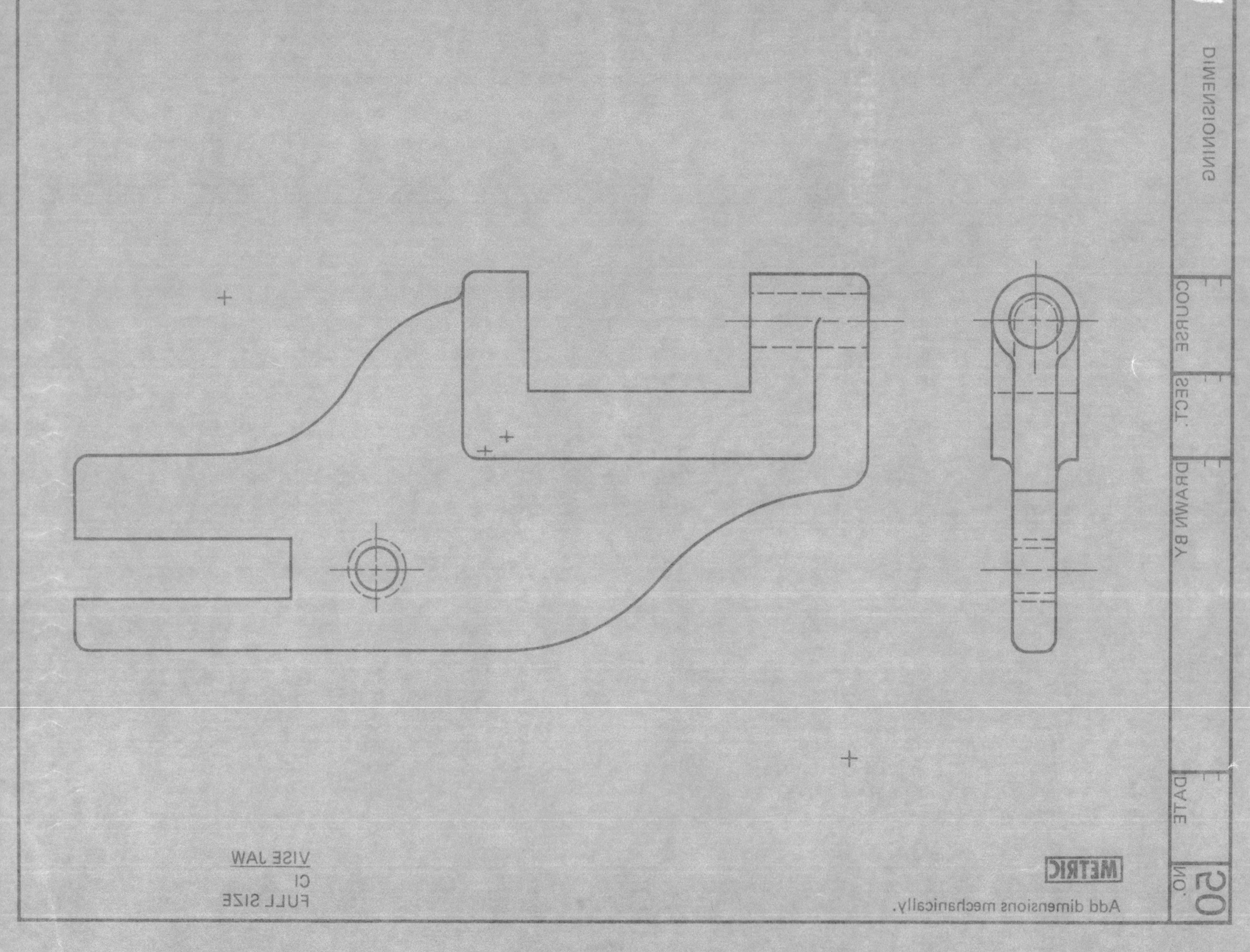

VISE JAW
CI
FULL SIZE
METRIC
Add dimensions mechanically.
DIMENSIONING
COURSE
SECT.
DRAWN BY
DATE
NO.
50

Draw the necessary views of the WASHER SEAL and add dimensions mechanically. Three place decimals have a tolerance of ± .010 except as noted. The remaining dimensions are two place decimals and have a tolerance of ± .03 (Dimensions B and C are three-place decimal dimensions.) Express all three place decimal values with a high and low limit. The WASHER SEAL is turned from 2 1/4 inch diameter cold drawn steel. All surfaces are finished.

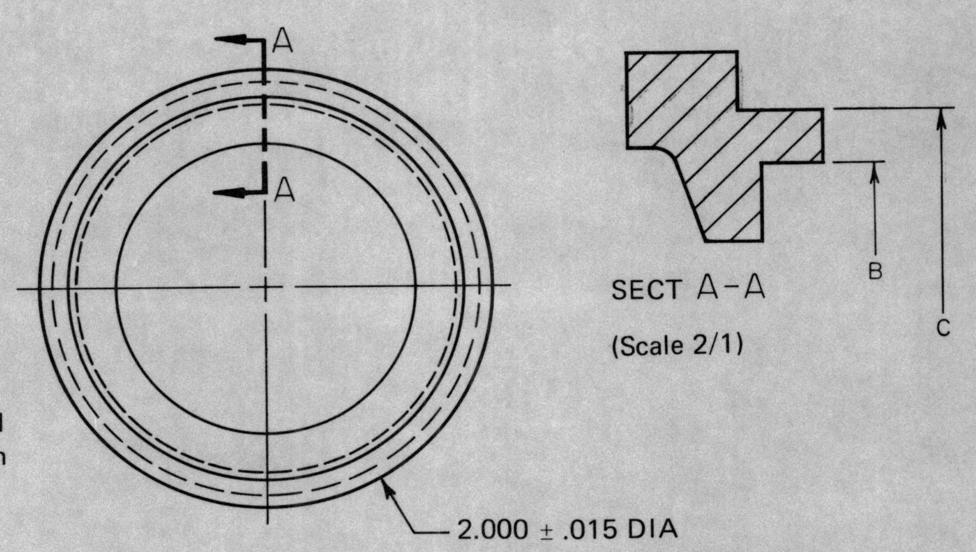

WASHER SEAL

(Not to scale)

DIMENSIONING	COURSE	SECT.	DRAWN BY	DATE	NO. 52

Draw the necessary views of the WASHER SEAL and add dimensions mechanically. Three place decimals have a tolerance of ± .010 except as noted. The remaining dimensions are two place decimals and have a tolerance of ± .03 (Dimensions B and C are three-place decimal dimensions.) Express all three place decimal values with a high and low limit. The WASHER SEAL is turned from 2 1/4 inch diameter cold drawn steel. All surfaces are finished.

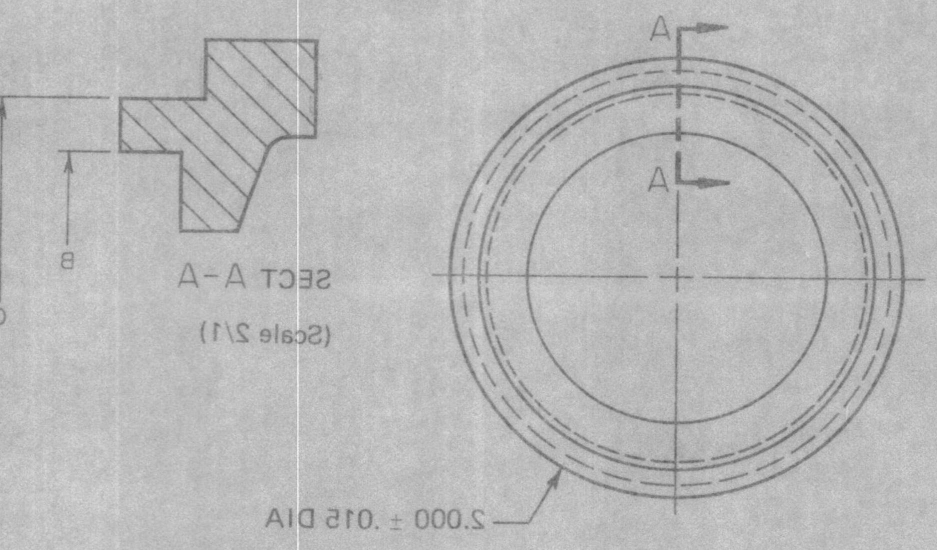

WASHER SEAL

(Not to scale)

DIMENSIONING	COURSE	SECT.	DRAWN BY	DATE	NO. 52

SHAFT EXTENSION COLLAR

Add dimensions to fully describe the object. Apply an RC4 fit to the 1.00 hole. Use broken out sections if necessary. Scale 1/2

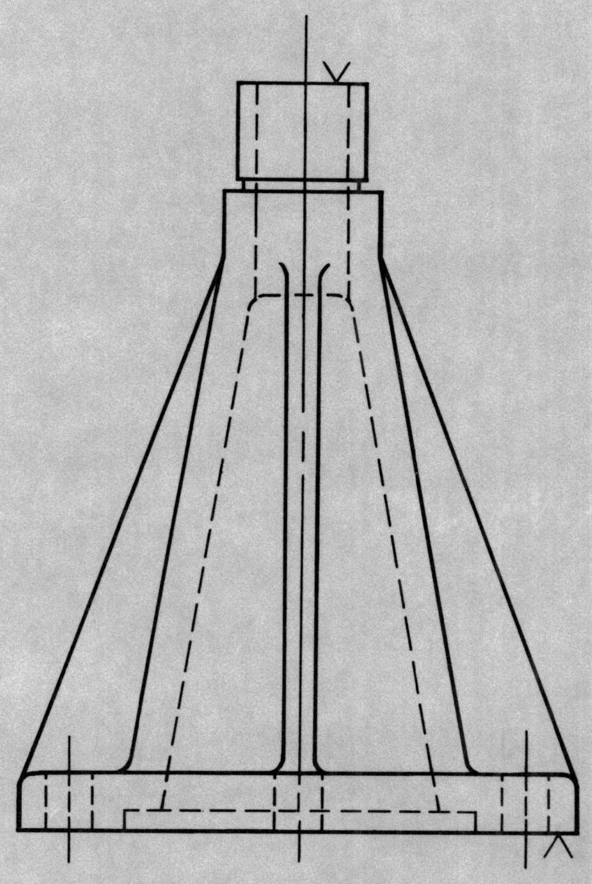

DIMENSIONING	COURSE	SECT.	DRAWN BY	DATE	NO. 54

SHAFT EXTENSION COLLAR

Add dimensions to fully describe the object. Apply an RC4 fit to the 1.00 hole. Use broken out sections if necessary. Scale 1/2

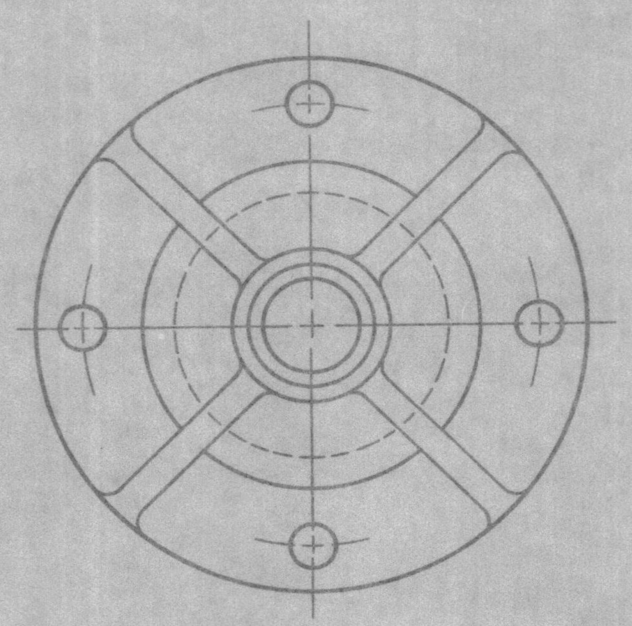

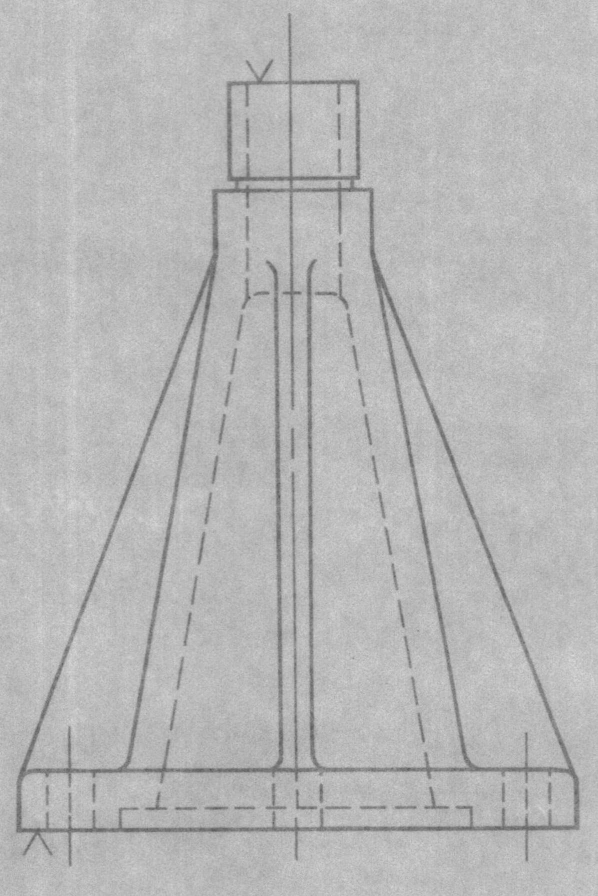

DIMENSIONING	COURSE	SECT.	DRAWN BY	DATE	NO. 54

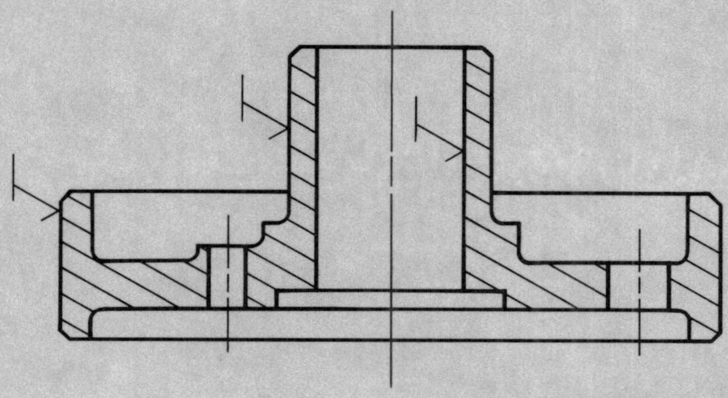

ALIGNMENT PLATE
CI
FULL SIZE

Add dimensions mechanically.

DIMENSIONING	COURSE	SECT.	DRAWN BY	DATE	NO. 55

ALIGNMENT PLATE

CI

FULL SIZE

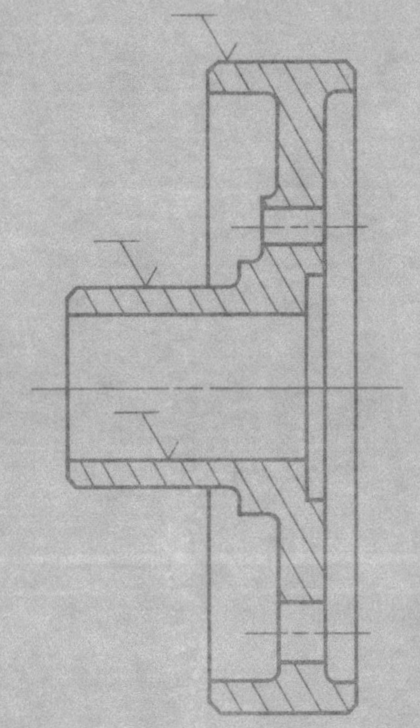

Add dimensions mechanically.

DIMENSIONING	COURSE	SECT.	DRAWN BY	DATE	NO. 55

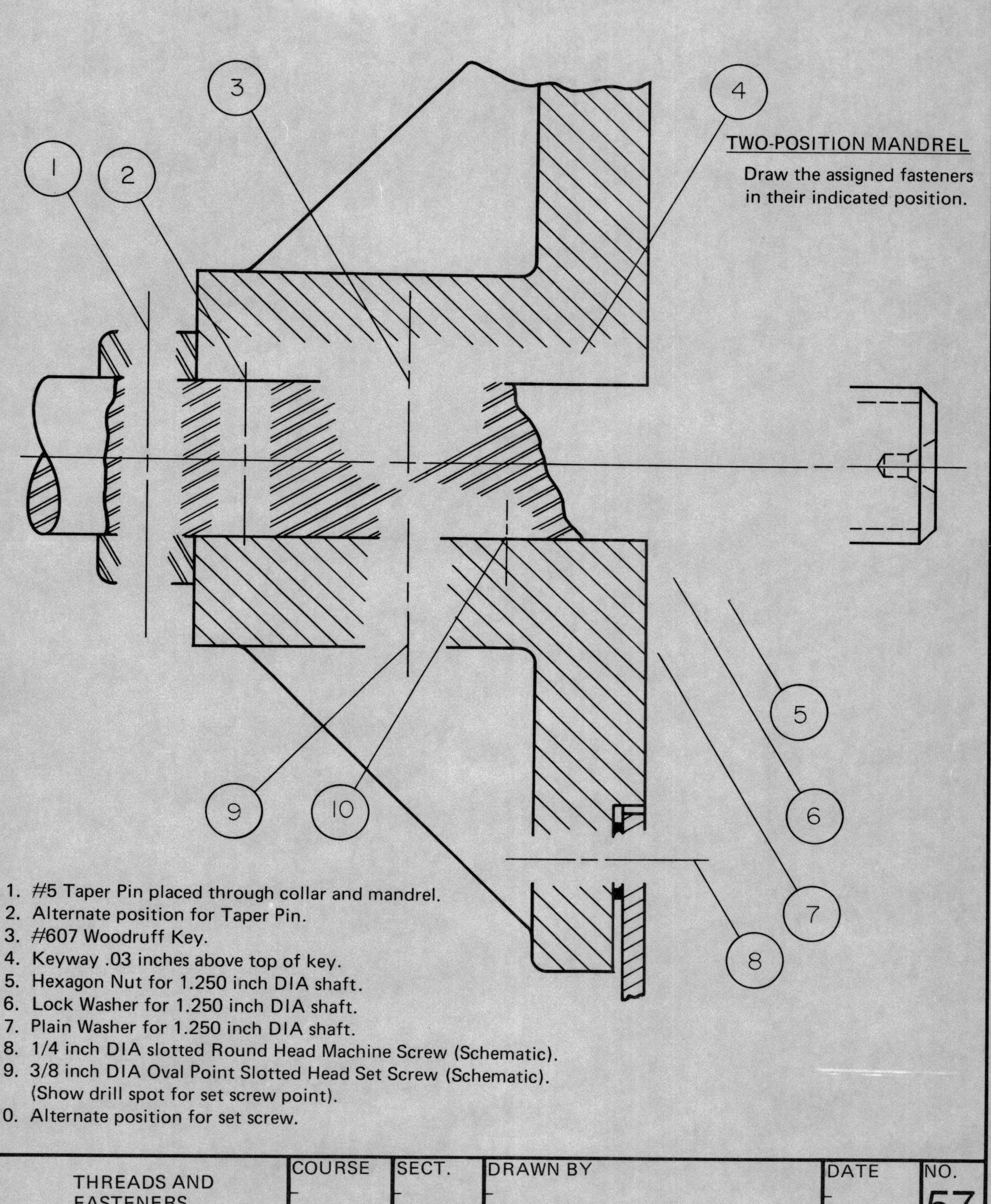

TWO-POSITION MANDREL

Draw the assigned fasteners in their indicated position.

1. #5 Taper Pin placed through collar and mandrel.
2. Alternate position for Taper Pin.
3. #607 Woodruff Key.
4. Keyway .03 inches above top of key.
5. Hexagon Nut for 1.250 inch DIA shaft.
6. Lock Washer for 1.250 inch DIA shaft.
7. Plain Washer for 1.250 inch DIA shaft.
8. 1/4 inch DIA slotted Round Head Machine Screw (Schematic).
9. 3/8 inch DIA Oval Point Slotted Head Set Screw (Schematic). (Show drill spot for set screw point).
10. Alternate position for set screw.

THREADS AND FASTENERS	COURSE	SECT.	DRAWN BY	DATE	NO. 57

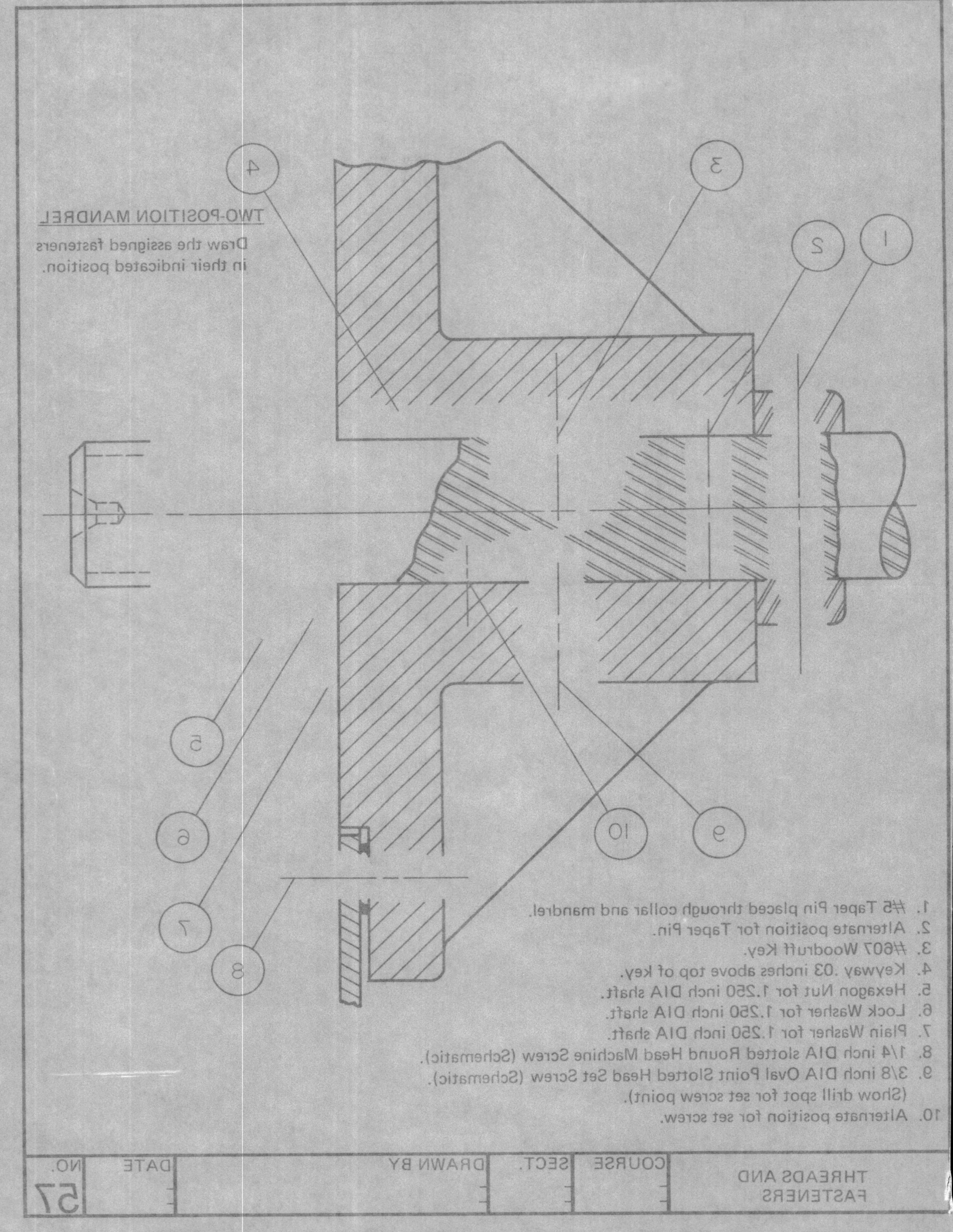

TWO-POSITION MANDREL
Draw the assigned fasteners in their indicated position.
1
2
3
4
5
6
7
8
9
10
1. #5 Taper Pin placed through collar and mandrel.
2. Alternate position for Taper Pin.
3. #607 Woodruff Key.
4. Keyway .03 inches above top of key.
5. Hexagon Nut for 1.250 inch DIA shaft.
6. Lock Washer for 1.250 inch DIA shaft.
7. Plain Washer for 1.250 inch DIA shaft.
8. 1/4 inch DIA slotted Round Head Machine Screw (Schematic).
9. 3/8 inch DIA Oval Point Slotted Head Set Screw (Schematic).
(Show drill spot for set screw point).
10. Alternate position for set screw.
THREADS AND FASTENERS
COURSE
SECT.
DRAWN BY
DATE
NO.
57

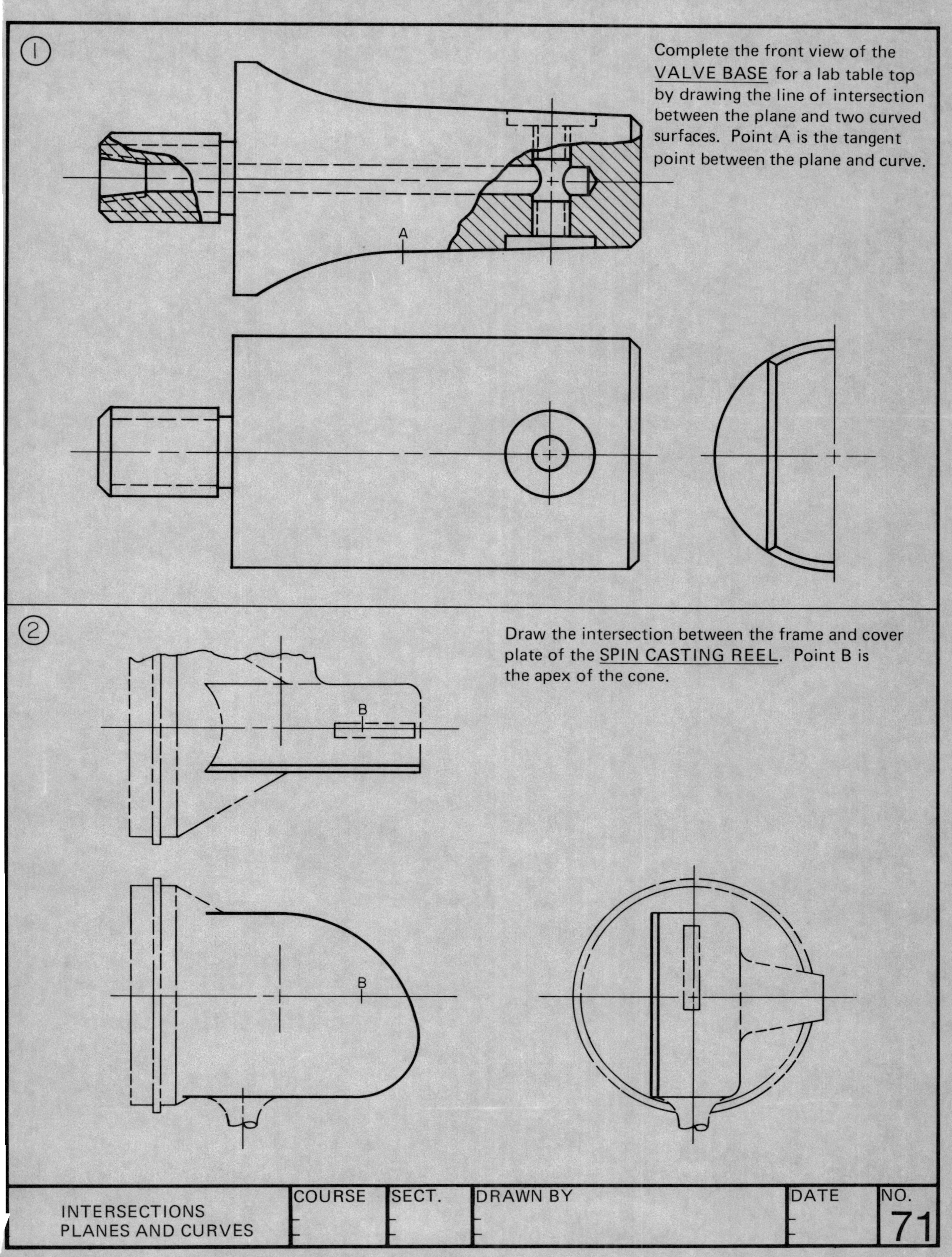

1
Complete the front view of the VALVE BASE for a lab table top by drawing the line of intersection between the plane and two curved surfaces. Point A is the tangent point between the plane and curve.
A
2
Draw the intersection between the frame and cover plate of the SPIN CASTING REEL. Point B is the apex of the cone.
B
B
INTERSECTIONS
PLANES AND CURVES
COURSE
SECT.
DRAWN BY
DATE
NO.
71

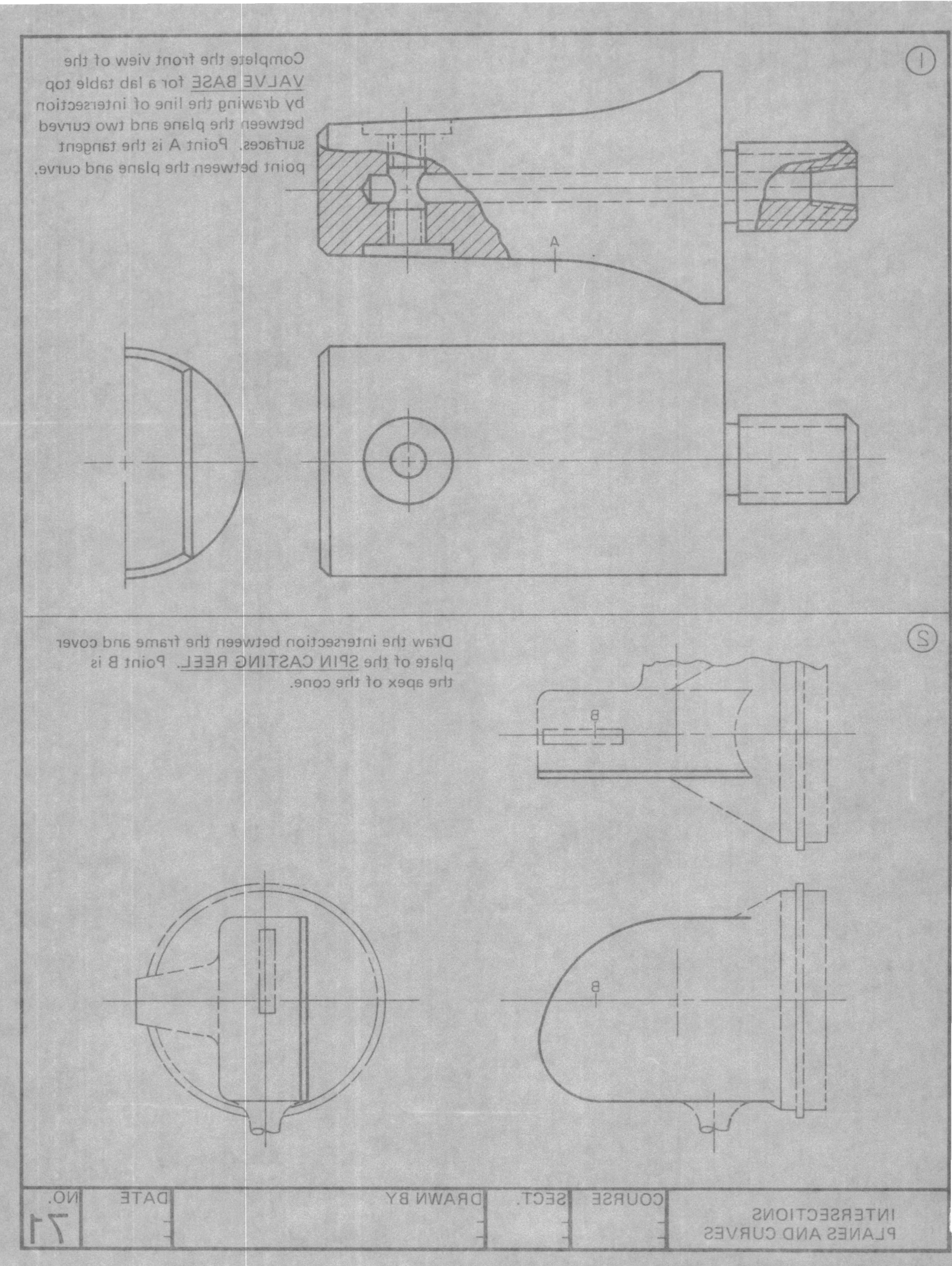

1
Complete the front view of the VALVE BASE for a lab table top by drawing the line of intersection between the plane and two curved surfaces. Point A is the tangent point between the plane and curve.
A
2
Draw the intersection between the frame and cover plate of the SPIN CASTING REEL. Point B is the apex of the cone.
B
B
INTERSECTIONS
PLANES AND CURVES
COURSE
SECT.
DRAWN BY
DATE
NO.
71

(1) Draw the line of intersection between the two prisms. Assume both prisms are open at the line of intersection. Show all constructions.

(2) Draw the line of intersection between the two unlike diameter cylinders. Assume that both cylinders are open at the line of intersection. Show all constructions.

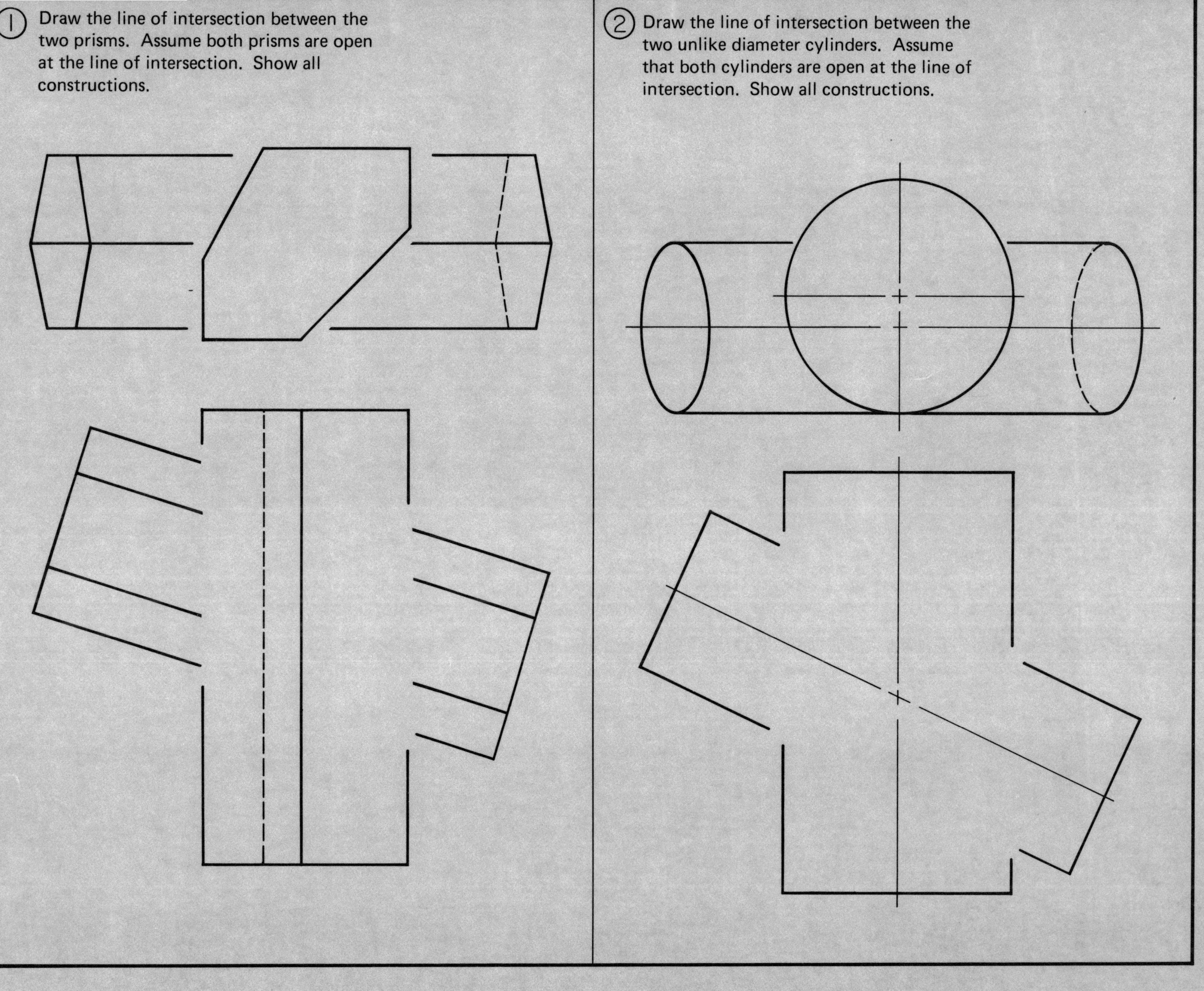

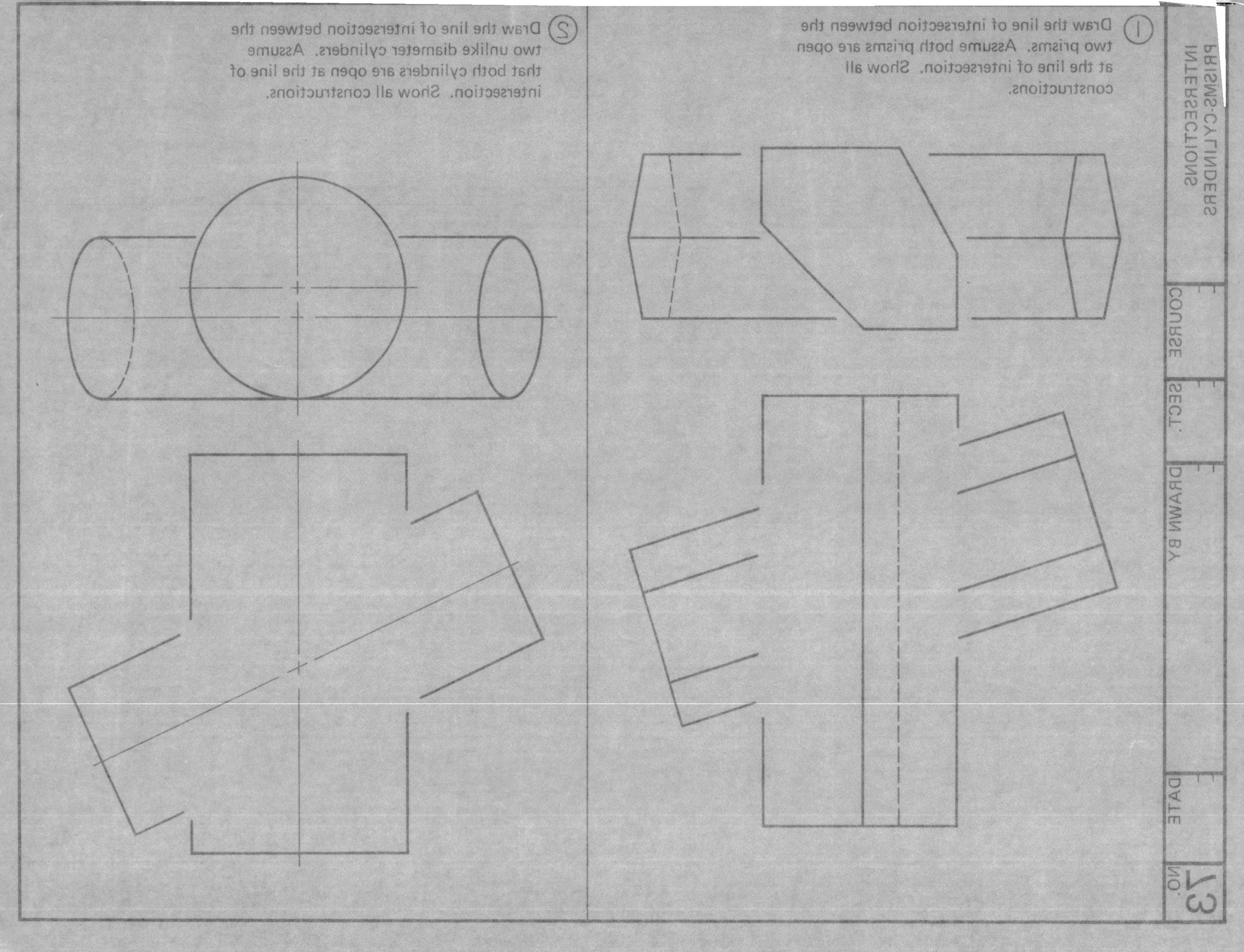

(1) Draw the line of intersection between the two prisms. Assume both prisms are open at the line of intersection. Show all constructions.

(2) Draw the line of intersection between the two unlike diameter cylinders. Assume that both cylinders are open at the line of intersection. Show all constructions.

1

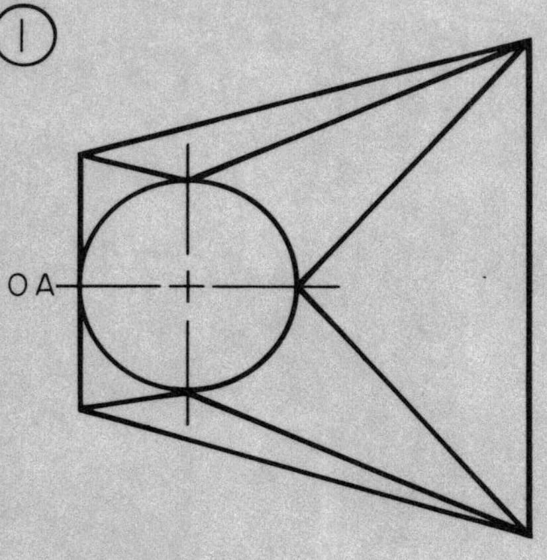

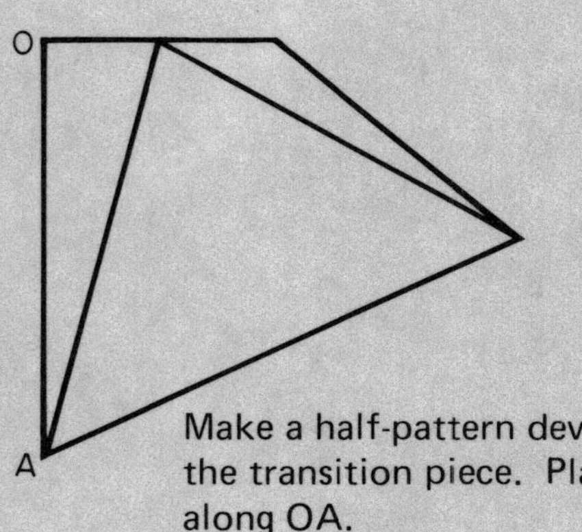

Make a half-pattern development of the transition piece. Place the seam along OA.

2 Develop the lateral surfaces of the transition piece. Place the seam along AX.

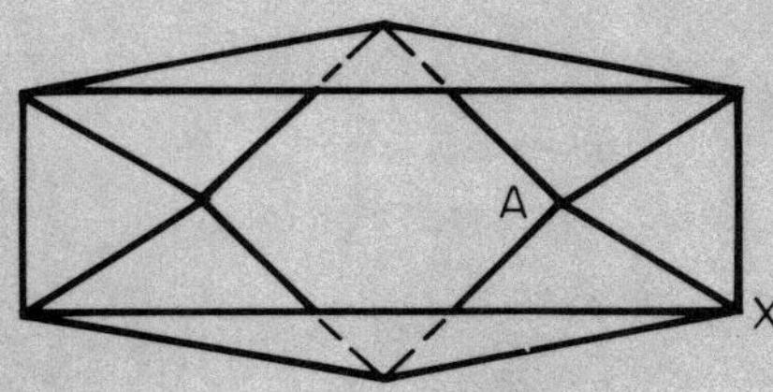

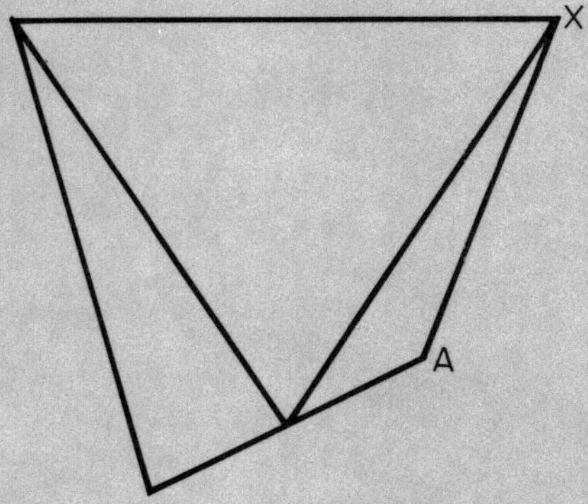

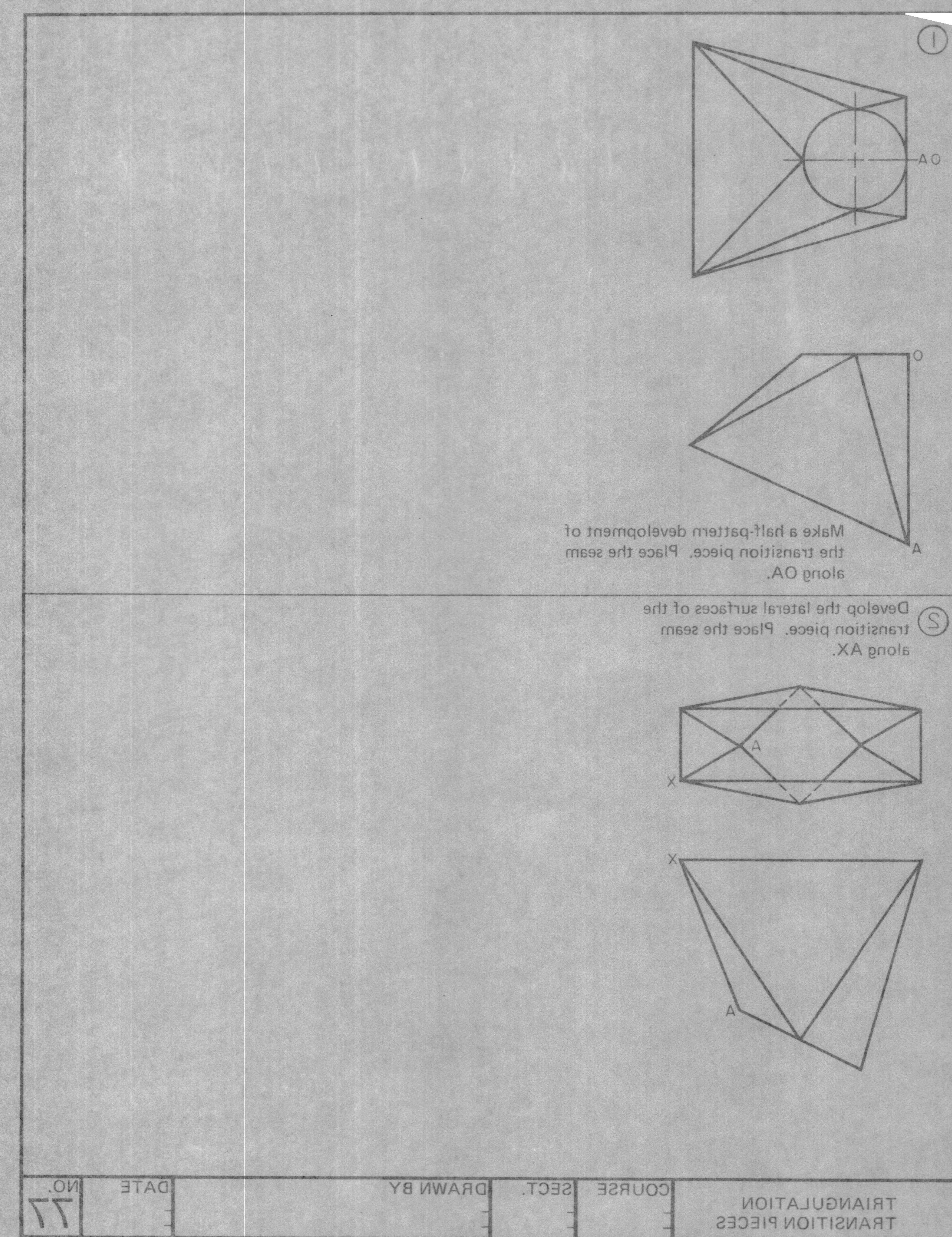

1
O A
O
A
Make a half-pattern development of the transition piece. Place the seam along OA.
2
Develop the lateral surfaces of the transition piece. Place the seam along AX.
A
X
X
A
TRIANGULATION
TRANSITION PIECES
COURSE
SECT.
DRAWN BY
DATE
NO.
77

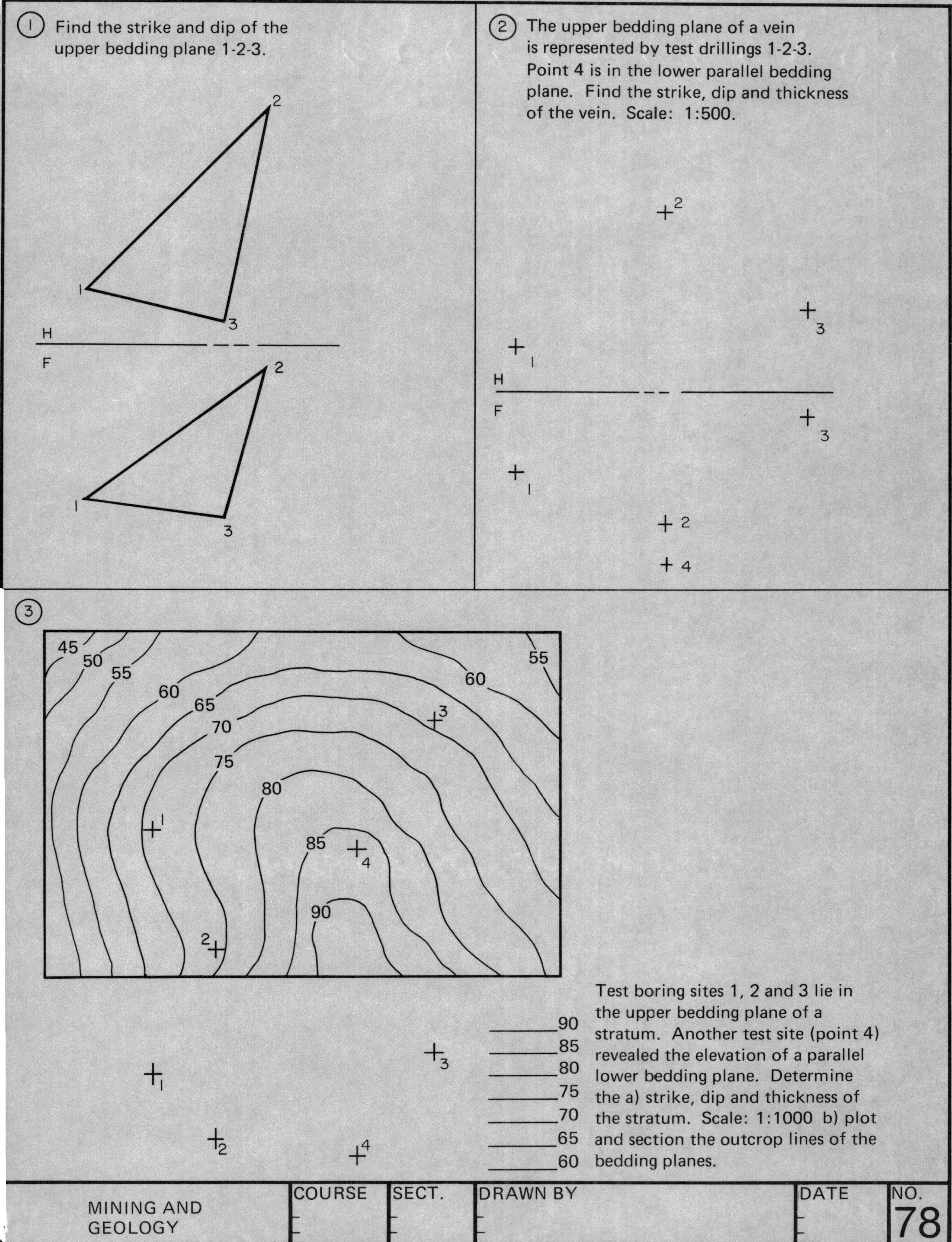

1 Find the strike and dip of the upper bedding plane 1-2-3.
2
1
3
H
F
2
1
3
2 The upper bedding plane of a vein is represented by test drillings 1-2-3. Point 4 is in the lower parallel bedding plane. Find the strike, dip and thickness of the vein. Scale: 1:500.
2
3
1
H
F
3
1
2
4
3
45
50
55
60
65
70
75
80
85
90
55
60
3
1
4
2
90
85
80
75
70
65
60
1
3
2
4
Test boring sites 1, 2 and 3 lie in the upper bedding plane of a stratum. Another test site (point 4) revealed the elevation of a parallel lower bedding plane. Determine the a) strike, dip and thickness of the stratum. Scale: 1:1000 b) plot and section the outcrop lines of the bedding planes.
MINING AND GEOLOGY
COURSE
SECT.
DRAWN BY
DATE
NO.
78

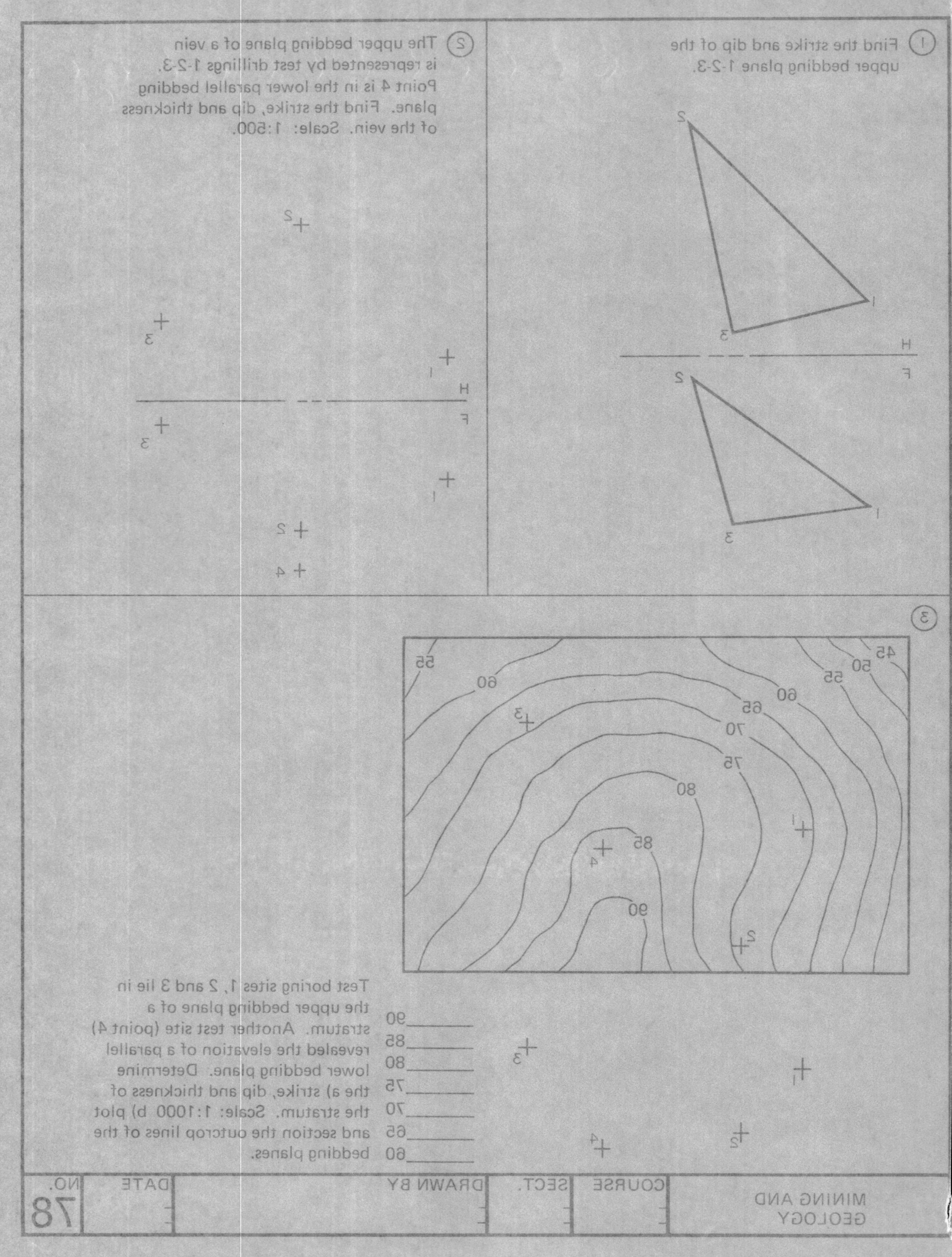
1
Find the strike and dip of the upper bedding plane 1-2-3.
2
1
3
H
F
2
1
3
2
The upper bedding plane of a vein is represented by test drillings 1-2-3. Point 4 is in the lower parallel bedding plane. Find the strike, dip and thickness of the vein. Scale: 1:500.
2
3
1
H
F
3
1
2
4
3
45
50
55
60
65
70
75
80
85
90
55
60
3
1
4
2
Test boring sites 1, 2 and 3 lie in the upper bedding plane of a stratum. Another test site (point 4) revealed the elevation of a parallel lower bedding plane. Determine the a) strike, dip and thickness of the stratum. Scale: 1:1000 b) plot and section the outcrop lines of the bedding planes.
90
85
80
75
70
65
60
3
1
2
4
MINING AND GEOLOGY
COURSE
SECT.
DRAWN BY
DATE
NO.
78

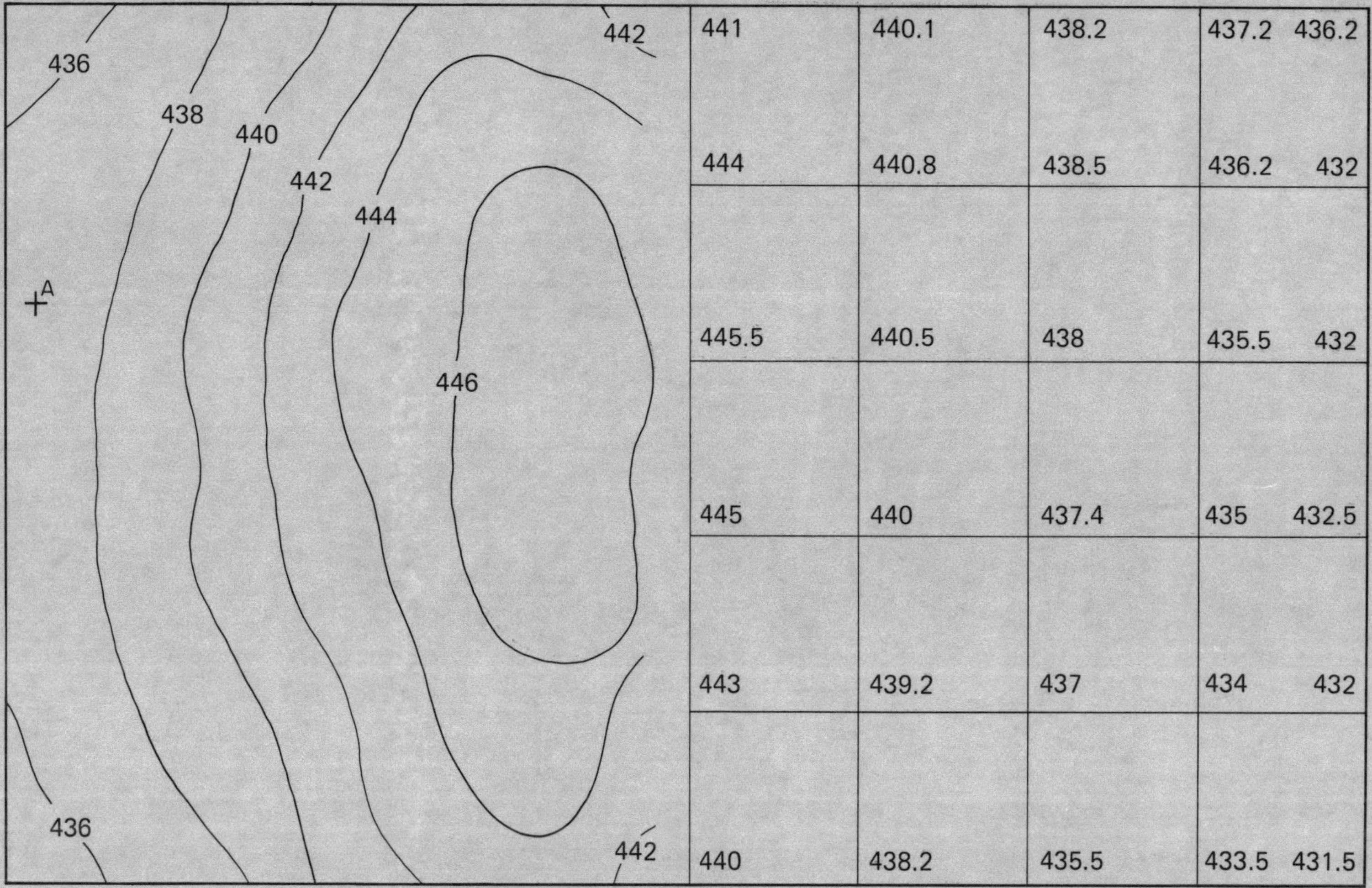
436
438
440
442
444
446
442
436
442
A
441
440.1
438.2
437.2
436.2
444
440.8
438.5
436.2
432
445.5
440.5
438
435.5
432
445
440
437.4
435
432.5
443
439.2
437
434
432
440
438.2
435.5
433.5
431.5

The centerline of a level road bears S 88° E from point A at an elevation of 438′. The road is 15.0′ wide. The road at its specified elevation will require cutting and filling. The cut and fill have a slope of 6/10 and 4/10 respectively. Complete the contour lines on the grid survey and a) locate and dimension the road and b) plot and identify the cut and fill areas and draw a profile at the maximum cut and maximum fill. Scale: 1″ = 10.0′

Profiles

CIVIL ENGINEERING
CUT AND FILL

COURSE	SECT.	DRAWN BY	DATE	NO.
				79

441	440.1	438.2	437.2	436.2
444	440.8	438.5	436.2	432
445.5	440.5	438	435.5	432
445	440	437.4	435	432.5
443	439.2	437	434	432
440	438.2	435.5	433.5	431.5

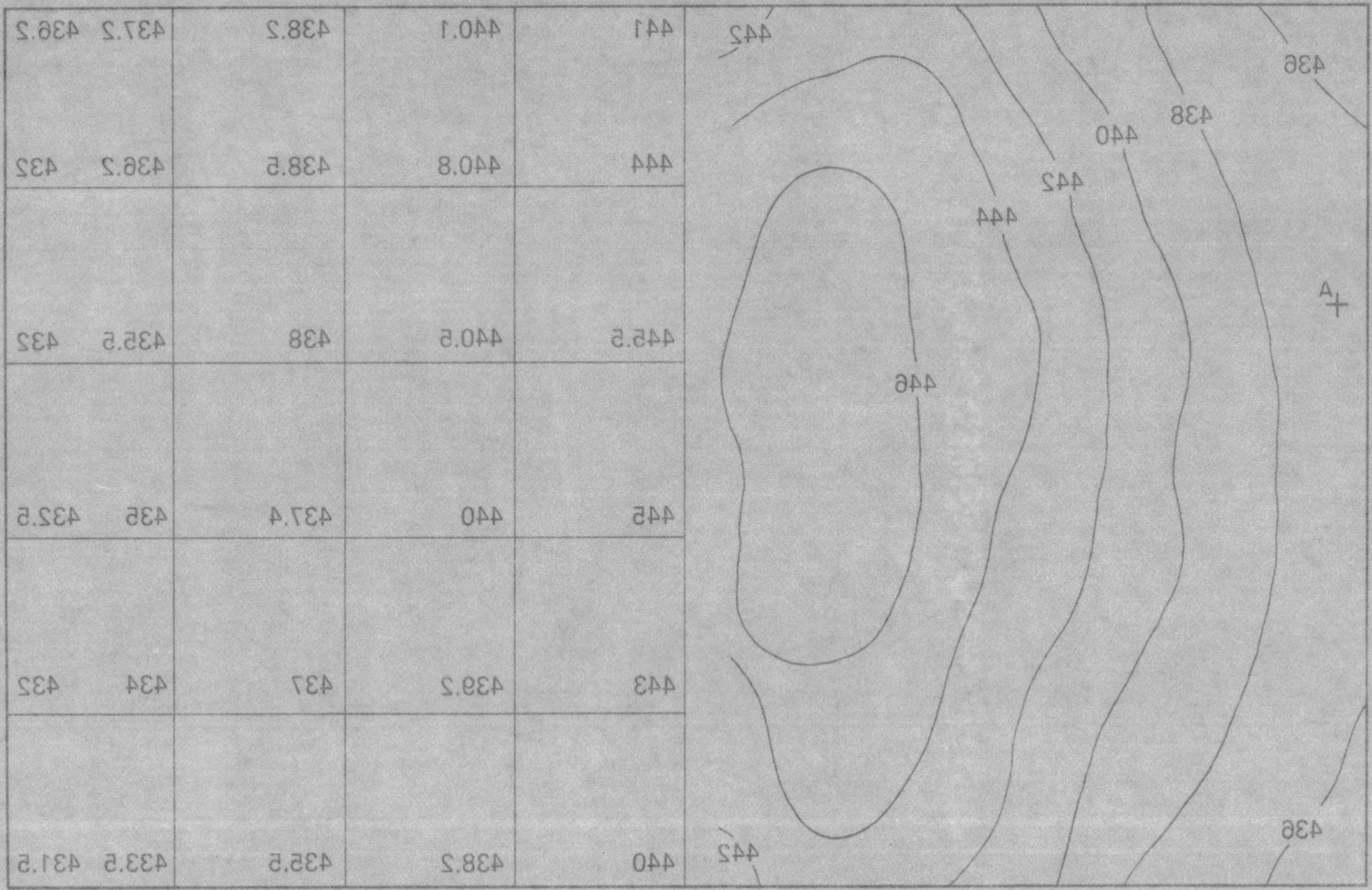

The centerline of a level road bears S 88° E from point A at an elevation of 438'. The road is 15.0' wide. The road at its specified elevation will require cutting and filling. The cut and fill have a slope of 6/10 and 4/10 respectively. Complete the contour lines on the grid survey and a) locate and dimension the road and b) plot and identify the cut and fill areas and draw a profile at the maximum cut and maximum fill. Scale: 1" = 10.0'

Profiles

CIVIL ENGINEERING
CUT AND FILL

COURSE	SECT.	DRAWN BY	DATE	NO.
				79

Establish the top and front views of the spherical triangle ABC. Angle B = 100° Side BA = 52° Side BC = 50°. Solve for angles A and C and side CA.
Scale: Full

Angle A= ______________

Angle C= ______________

T.L. of CA= ______________

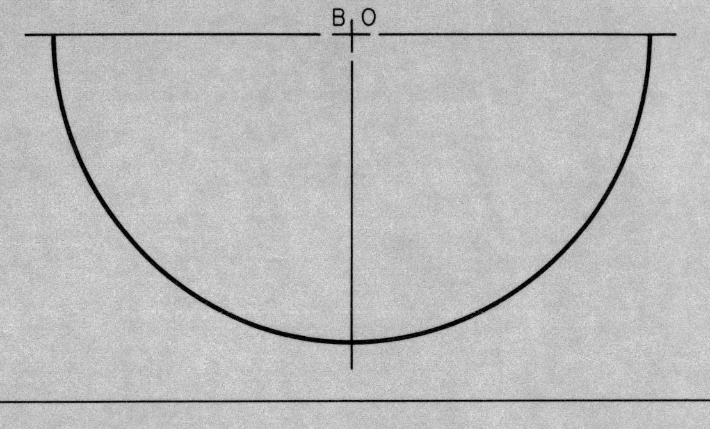

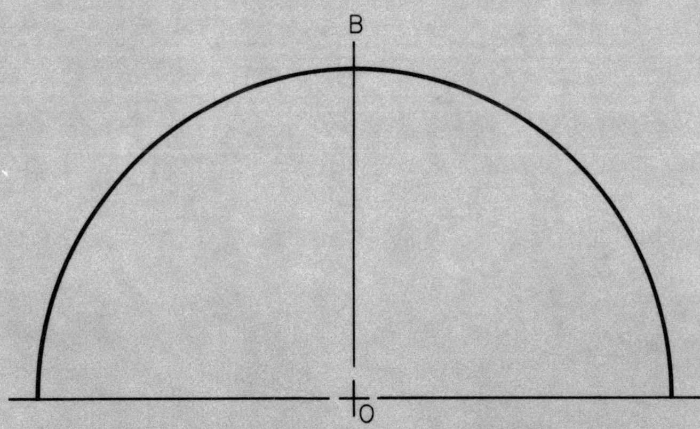

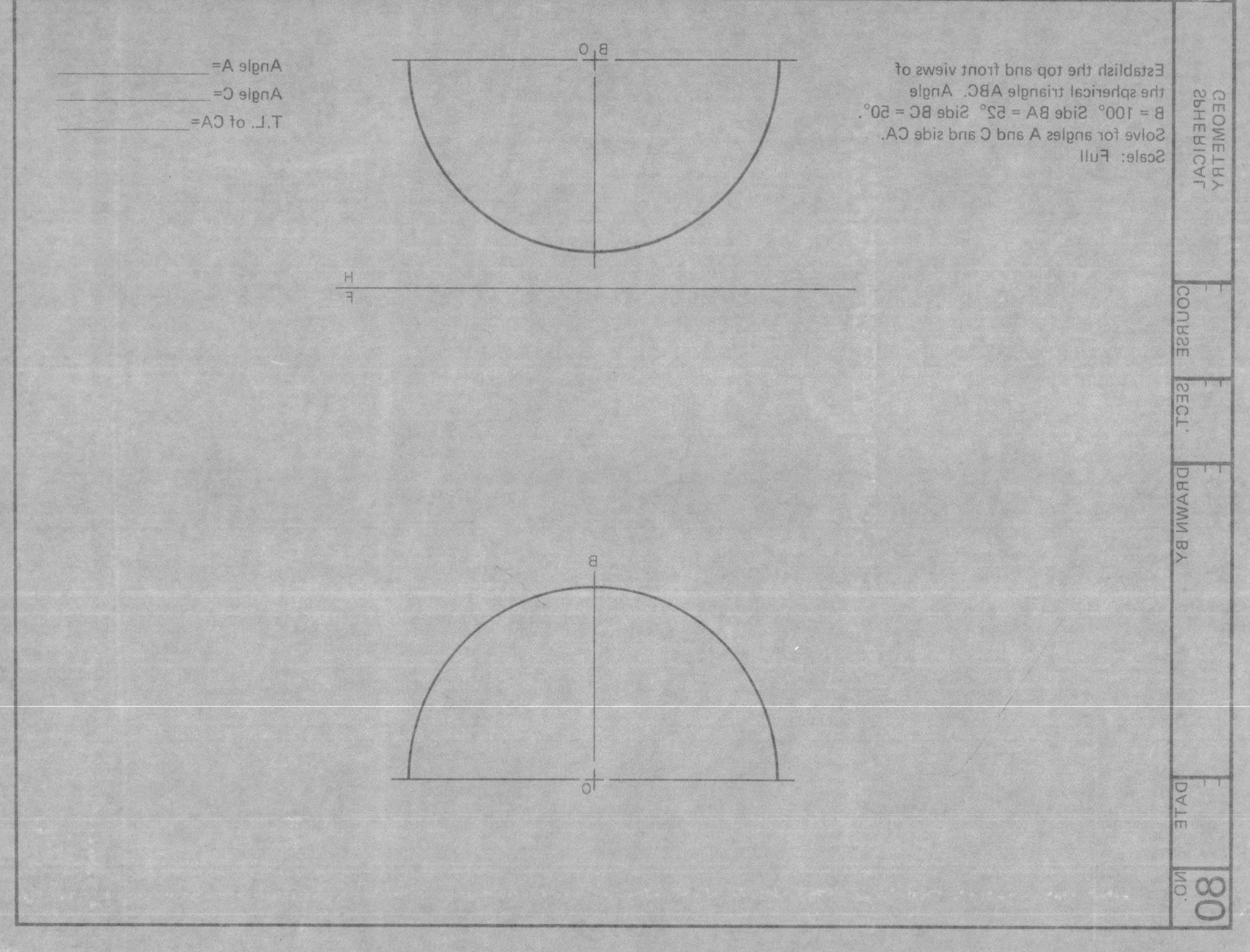

Establish the top and front views of the spherical triangle ABC. Angle B = 100° Side BA = 52° Side BC = 50°. Solve for angles A and C and side CA. Scale: Full
Angle A= ____
Angle C= ____
T.L. of CA= ____
B
O
H
F
B
O
SPHERICAL GEOMETRY
COURSE
SECT.
DRAWN BY
DATE
NO.
80

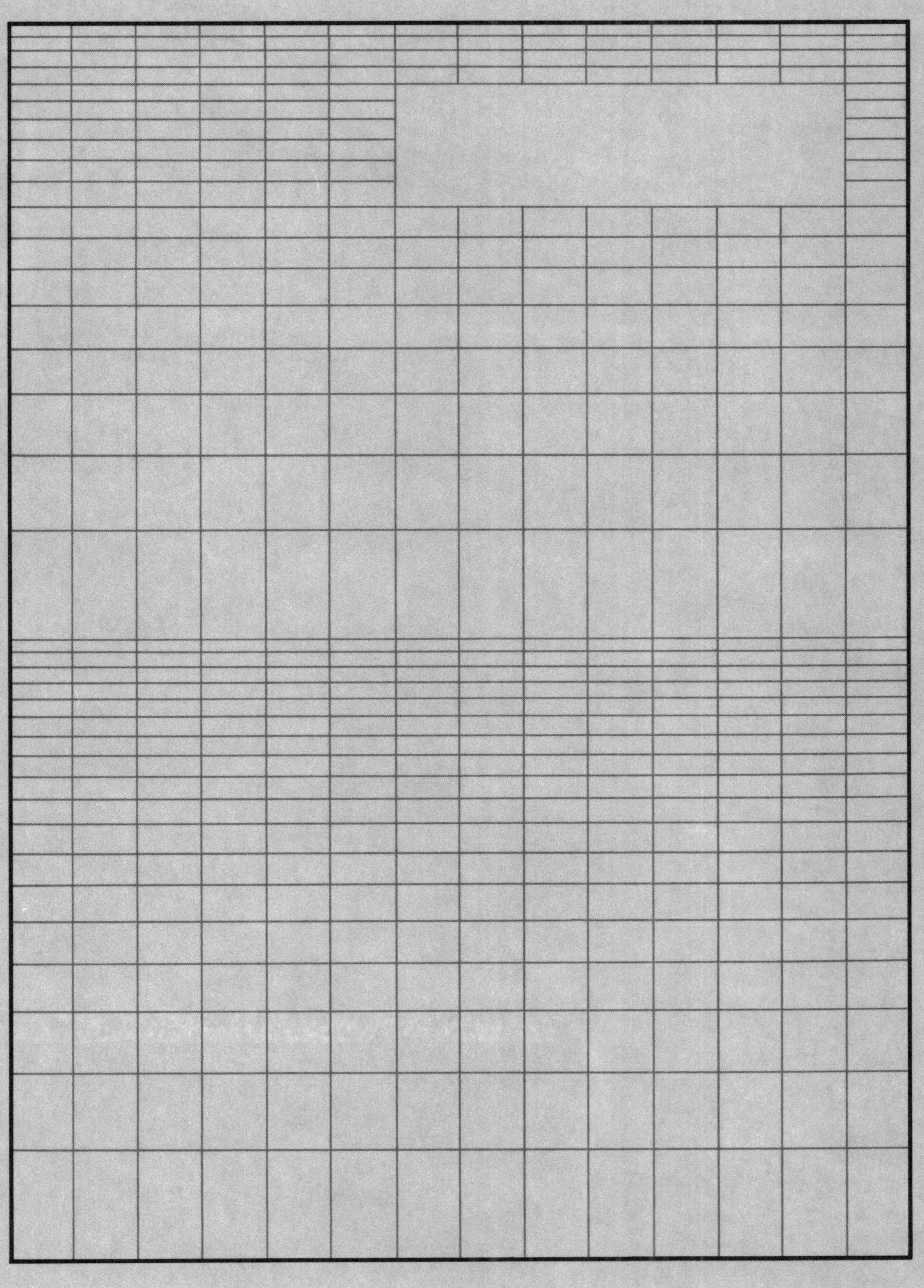

SEMI-LOGARITHMIC COORDINATE GRAPH	COURSE	SECT.	DRAWN BY	DATE	NO. 86

NO.	DATE	DRAWN BY	SECT.	COURSE	SEMI-LOGARITHMIC COORDINATE GRAPH
86					

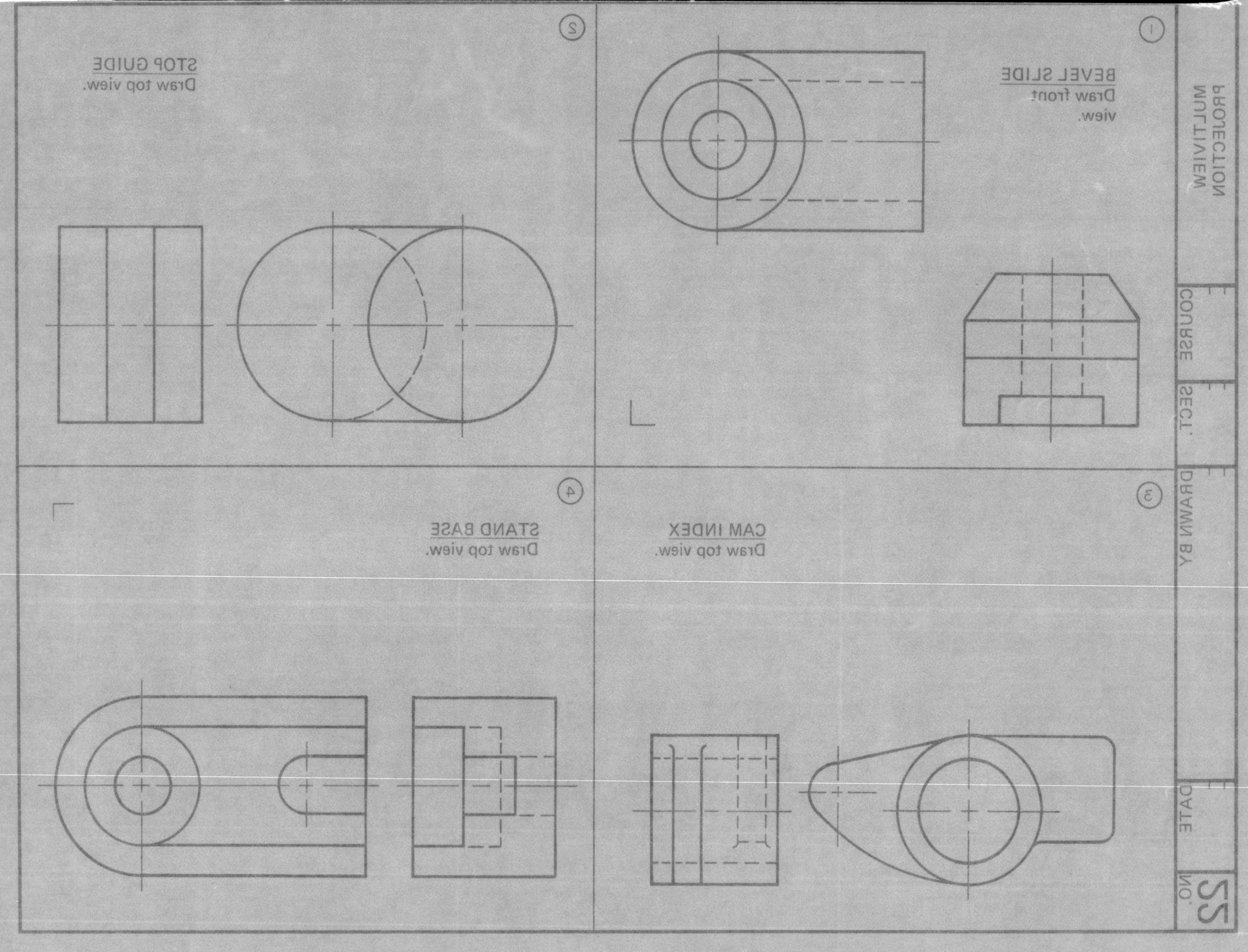

1
BEVEL SLIDE
Draw front view.
2
STOP GUIDE
Draw top view.
3
CAM INDEX
Draw top view.
4
STAND BASE
Draw top view.
MULTIVIEW PROJECTION
COURSE
SECT.
DRAWN BY
DATE
NO. 22